Climates of Hunger

Mankind and the World's Changing Weather

CLIMATES OF HUNGER

MANKIND

AND

THE WORLD'S CHANGING WEATHER

Reid A. Bryson
and
Thomas J. Murray

THE UNIVERSITY OF WISCONSIN PRESS

Published 1977
The University of Wisconsin Press
Box 1379, Madison, Wisconsin 53701

The University of Wisconsin Press, Ltd.
70 Great Russell Street, London

First printing

Printed in the United States of America

For LC CIP information see the colophon

ISBN 0-299-07370-X

Contents

Illustrations

Foreword

I am quickly turned off by books written by specialists for other specialists—which this book is not. It is a book I shall cherish—and persuade my students to read. Climate has been one of the many historic influences that historians have failed to take seriously. Or in some cases they have taken it too seriously, while lacking the factual basis to back up their theories. Most climatologists don't read history, so that they, too, haven't been able to make the connection.

Bryson and Murray belong to a small but growing group of generalists who have technical training *and* a grasp of the facts of human history. They are able to make the connection. The fall of Mycenae is plausibly related to sudden drought. So is the southward migration of the Hittites. So is the sad end of Harappa, aided in this case by their own foolish use of the land. And so are many other historic changes where we have hitherto leaned on institutional failings, hubris, or sheer chance to explain events. Many will quarrel with the connections offered. But none will have any trouble understanding. The authors have written for the citizen, not a narcissistic coterie.

They are not alone. In the last few years a small but precious breed of climatic historians has transformed our knowledge of the past thousand years: Páll Bergthorsson, patiently reconstructing the record of sea ice around the shores of Iceland, the real bell-wether of northern hemisphere climate; Gordon Manley, calibrating the old thermometers of England to give us a temperature record back to the seventeenth century; Hubert Lamb, showing that one can draw tentative weather maps long before the days of weather stations; and Le Roy Ladurie, combing the guard-books of monasteries and vineyards, using the grape and its productivity as a surrogate for instruments. These, and many more, have begun to show what climate *was*—and may be again.

Nothing could be more timely, for climate is once again at the heart of the matter. The 1970s have produced some spectacular droughts, floods, gales, and freeze-ups—many more than in recent decades. These things have hit us in our pocketbooks. They have caused crop failures, and hence price inflation. They have reduced some of Africa's new republics to the point of despair and bankruptcy. And they have raised again the specter of the 1930s, that our own wheat fields may again blow away, just when the world needs them most.

I hope that this little book will make a dent in the armor of complacency that protects our politicians against this new intransigence on nature's path. God knows we need awareness—and the number of times my phone rings shows that the awareness exists among the media and the public. But we also need information, and a perspective. We need to be able to say: these things that we predict have happened before, and can therefore happen again. We need critical, self-aware prophets. Bryson and Murray are clearly among them.

F. Kenneth Hare

The Institute for Environmental Studies
University of Toronto

December 15, 1976

Prologue

In 1973 an international group of scientists wrote to the President of the United States about a matter of grave concern. They were specialists in the history of ice ages, and they could see from the rhythm of past ice ages the possibility of another ice age within centuries, and almost positively within a few millennia. There is no imminent danger of a new continental ice sheet like the one that existed more than 10,000 years ago from the Arctic to the Great Lakes and the Atlantic to the Pacific. These take millennia to grow—although we should not ignore changes on that time scale, because we are building dams and nuclear waste facilities with planned lifetimes of centuries and longer.

But climatic changes preceding the formation of such ice sheets are also extremely important, as are the fluctuations that occur throughout unglaciated times.

For three decades the nations of the world, aided by superb agricultural science and technology, have kept food production abreast of a rapidly growing population. In the late 1960s it appeared that the "Green Revolution" was gaining in the constant race against famine. But since 1972 we have seen the international economy—as well as world news—strongly influenced by climatic events.

The Soviet Union purchased 18,000,000 tons of grain from the United States in 1972 and another 12,000,000 tons in 1975, with repercussions on domestic food prices and on our balance of payments. The Russian purchases were directly related to drought. In 1974 United States corn production was significantly reduced as a result of drought, frost, and in some cases overly wet weather.

During these years we have seen drought in the Sahel and in Central America, severe frost damage to Brazilian coffee, poor monsoons in South Asia, and many other climatic vicissitudes. Moreover, the winter of 1977, with record cold in the midwestern and eastern United States, snow in southern Florida, and a continuing drought in the Great Plains and the West, emphasized again the effects of climate on human activity.

And yet, in a world where grain yields have diminished since 1972, reserves of grain are very small, and the costs of producing and buying food have sharply risen—partly as a result of increased petroleum prices—human population continues to grow. If the world is to feed seven billion people by the end of the century it cannot afford more production years like those since 1972. Every year in which the climate intervenes in food production is serious. Climatic fluctuations become far more important than when there are large surpluses and rising production.

Why "climatic fluctuations" rather than "years with adverse climate"? Many of our food crops are very well adapted to the climate under which they are grown—for example, corn. This means that the best yield is obtained under climatic conditions like those of the recent past (often called "normal" climate). Any departure from "normal" brings a lower yield; any fluctuation is adverse for such well-adapted crops.

We have not yet resolved the question of how soon the next ice age will begin, or even whether we are already in the transition period. However, we are not left without clues about what is possible.

This book sketches some climatic changes of the past and their effects on civilizations. Climatic history shows us that:

1. We should not think of the climate and weather of our lifetimes as an unchanging "normal." The mid-twentieth century is not typical of the previous thousand years—and it is even less typical of the past one million years. We have ignored what climatic change can do to our agriculture, and our populations. We have depleted the petroleum resources we might need for massive readjustments when climate reverts to a more typical pattern.
2. Climate not only varies year by year but can change rapidly to a new multi-year average.
3. The climate, once changed, can stay changed for long periods of time. Within the last thousand years there has been a 200-year drought in the United States corn and spring wheat belt.

4. Records of past climates also show that those times in earth history when temperatures in high latitudes have been lower, have generally been times of greater weather extremes and erratic or absent monsoon rains in Asia. These would affect United States domestic and foreign policy as well as the people in those lands.

   From about 1945 to the present, *temperatures at high latitudes have fallen irregularly*.
5. Once an ice age ends, the climate usually stays interglacial for ten to twelve thousand years. The present interglacial is about 10,800 years old.

We cannot predict with certainty the shape of the climatic future. But then decision-makers rarely have the advantage of precise foresight. Any policy is an effort to make the most of the future, with all its unknowns.

Just as an insurance policy is a recognition of risks and contingencies, so our national policy must recognize certain climatic risks and contingent problems.

Even climatic fluctuations that appear to be small in size can be significant economically. Our research at the University of Wisconsin-Madison shows that an increase of 1°C (1.8°F) in the summer temperatures in the northern plains can reduce the gross dollar income of the spring wheat farmers by $131 million, and a modest 20 percent shortfall of precipitation can cost another $137 million. Climatic variation, like death and taxes, is certain. We know of no century with constant climate. But a more important question is whether the climate will be approximately like that to which our activities have been adapted or whether we will be dealing with a major climatic shift.

> The probability of the next decade being less favorable than the 1956–73 period in the American corn belt is very high—around 98 chances in 100, if past history is a guide.
>
> Considering only the time since 1880, three out of four decades were colder than the 1931–60 average (for the Northern Hemisphere).
>
> Since A.D. 1600, 95 percent of the decades in the far northern Atlantic have been colder than the 1931–60 average.

These facts and probabilities indicate the likelihood of a more unstable climate than that of the decades just prior to 1970, and of a higher number of poor monsoons in Asia, and poor agricultural years in North America, Europe, and the USSR. While scientists do not agree on all the above points, most of those who have analyzed the data agree that the next decade will not be optimal for agriculture.

As long as world food grain reserves are low, as they are, and population is rising, as it is, there is danger for the nation in the political instabilities that climatically induced crop failures can and do produce. The American people also have a tradition of concern for those people who are suffering hardships. For these reasons, climatic variation must be considered as a factor in national policy formulation.

ACKNOWLEDGMENTS

There are a number of people who have made important contributions to the preparation of this book: John Ross started the authors on this project; Elizabeth Steinberg and Jan Blakeslee edited the manuscript and significantly improved it; Vicki Lant drew the maps and charts; Stephen Schneider reviewed the manuscript and suggested changes; Pam Egan Johannsen and Vicki Lant drew the chapter heading sketches; Donald Vincent worked on references and indexing; Barbara Abbott made suggestions on an early draft.

But most of all the book exists because of those who had faith over the years of research.

I TWO TALES OF FAMINE

CHAPTER I

A Drought in Ancient Greece

CLIMATES OF HUNGER are changed climates, climates that no longer support the crops and herds, berries, fruits, and game they once did. Climates change: a culture closely tied to a particular climate finds itself in danger.

Where agriculture is hard pressed to support a population, that population is in jeopardy—for climate does change, rapidly and significantly enough to alter the productivity of the land. Favorable climates that aid agriculture allow populations to grow beyond what later, less advantageous climates will tolerate.

We know climate can change, because climates have changed. Over millions of years, ice ages have come and gone—seven of them in the last million years alone. Even in the 10,000 years since the last ice age, climates have set boundaries for human activities. Human technology affects these boundaries, but so does changing climate. Cultures have developed and expanded, then withered and sometimes disappeared as new climates modified the potential of the land.

In our world today, as in the past, there are climates of hunger. As climate changes there will be both regions of improvement and regions of deterioration. Will lands now highly productive become too arid or too cold? This question is not fully answerable, but neither is it academic. Attempts to find answers, and to prepare for the climatic changes ahead, are vital to millions of people in the world.

In this book we will consider a number of past climatic changes, for the past is the key to the future.

On a sunny plain 60 miles southwest of Athens lie the ruins of the city of Mycenae. Twelve hundred years and more before the birth of Christ, Mycenae was the hub of a great civilization. Its massive main gate, with two stone lions on guard, its main walls, half a mile long and up to 30 feet thick, testify to the power it held. Its excavated tombs have revealed a wealthy and sophisticated warrior civilization with a farflung trade that dominated the Aegean and much of the Mediterranean seas for centuries.

Quite abruptly, before 1200 B.C., Mycenaean power began to decline. In 1230 B.C., the main palaces and granaries of Mycenae itself were attacked and burned. Other Mycenaean centers, including Pylos and Tiryns, also show signs of decay and destruction, but it is not known whether they were victims of their ties to a weakening Mycenae.

The decline and fall of Mycenaean civilization was so sudden and so complete that its memory survived only in the legends of Agamemnon and Achilles, of the fall of Troy and the voyages of Odysseus, given expression some 600 years later in the poems of Homer. And they remained legends until a stubborn amateur archaeologist with a love of Homer, Heinrich Schliemann, began to dig in the 1870s.

Neither then, nor now, has anyone fully explained the downfall of this vital culture.

Invaders?

Perhaps the most widely held current theory is that Dorian Greeks from the north overran Mycenaean Greece. In such a rapid fall of a major civilization, invasion is an obvious possibility. And the

Dorian dialect predominated in that part of Greece–the Peloponnesus–in classical times hundreds of years later. That explanation, however, has some problems. In a brief but incisively argued book, *Discontinuity in Greek Civilization* (1968), the distinguished classical scholar Rhys Carpenter considered the arguments for an invasion and found them wanting. Dorians, he points out, did not occupy Mycenae until at least two or three generations after the Mycenaean decline. In fact, the few people who continued there until Dorian times seem to have been of Mycenaean culture. But the most difficult problem appears to be the question of what route invaders could have taken. Invasion from the eastern Mediterranean is unlikely, since the islands of the South Aegean Sea, right in that line of approach, remained unmolested. The same holds true for invasion from the west (from Italy), or from the Adriatic coast to the north. Indeed, the island of Cephalonia, straddling any such invasion route, became a refuge for people from Greek regions to the east. To the northeast, Athens remained uninvaded; the region north of Mycenae became another refuge for Greeks of the Peloponnesus. The population increased there as well as on Cephalonia (see figure 1.1).

In short, Mycenae was ringed by routes invaders did *not* take, or at least left no sign of using. Historians have wrestled with this problem, and one of the most recent and most thorough, Vincent Desborough, reached the unsatisfactory conclusion that "the invaders did not settle in any areas they overran," but departed. Carpenter was prepared to offer a more radical solution: *"There were no invaders"* (1968, p. 39). What actually happened, he argued, was "not an invasion from outside, but an evacuation from within; not an enemy incursion, but a dispersal–a diaspora–of the Mycenaean inhabitants of the Peloponnese" (p. 41). And the cause of this dispersal? Drought, says Carpenter. He deduces, from the scanty evidence, large-scale movements of starving populations throughout the Aegean, and graphically depicts a final resort to violence by a drought-stricken populace. Mycenaeans themselves, not foreign invaders, burned the palaces and granaries of their overlords.

There is, Carpenter points out, evidence from classical Greek literature. Plato, in the *Timaeus*, relates a story of Solon, the great Athenian lawgiver, who had visited Egypt and was told by an Egyptian priest:

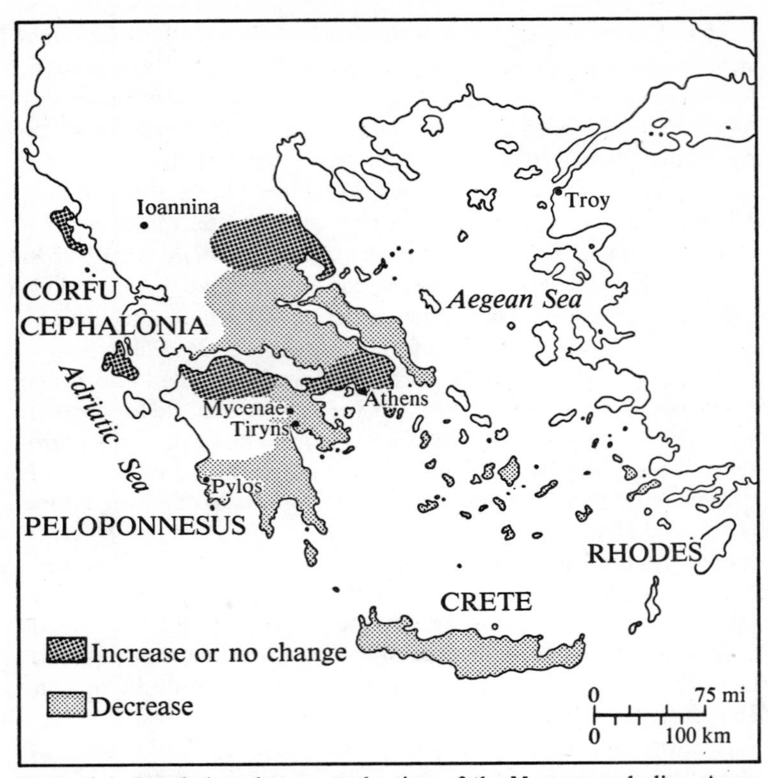

Figure 1.1. Population changes at the time of the Mycenaean decline. Areas where population changes are not central to the Mycenaean discussion or are not known are unshaded. After Donley, 1971, p. 4; based on descriptions in Carpenter, 1968, pp. 27–53.

"There have been many occasions of human destruction, and there will be many more. The chiefest source of these is fire and water; but countless lesser ones have been due to other causes. The legend that is told among you also, how Phaethon, child of the Sun, once yoked his father's car but, being unable to guide it on its wonted course, burnt up the earth beneath him and was himself destroyed by lightning—all this, though told in mythic guise, is true, inasmuch as a deviation of the celestial bodies moving past the earth does, at long intervals, cause destruction of earthly things through burning heat. . . ." [Carpenter, 1968, pp. ix–x]

The "fire and water" that destroyed Greeks from time to time, Carpenter says, were drought and flood (1968, p. x). One should not dismiss this story "told in mythic guise" as mere poetry. Schliemann, after all, did follow the myths of Homer to the sites of Troy and Mycenae.

Answers from climatology?

But Carpenter had to account for the fact that if drought struck Mycenae, it apparently did not strike Athens and the other nearby places where population did not decrease. For this question, Carpenter went beyond the literary record to consider possibilities of climate. He examined evidence of the rise and fall of water levels in the Mediterranean, and the pattern of westerly winds and storm tracks that bring rain to Greece.

Though not a climatologist, Carpenter pointed out that rain falls unevenly in mountainous country:

> After picking up moisture from the sea, [storms] tend to discharge it again in the form of rain on striking any relatively high transverse mountain barrier. Having shed part of their moisture load on [slopes facing them], they pass without further discharge across the lower lying country behind the mountain barrier. [Carpenter, 1968, p. 62]

Carpenter's observations are familiar to anyone who has spent time near mountains. In any mountainous region, the pattern of rainfall is uneven and derives in part from the wind direction and local topography. So in the mountainous land of Greece a change in the direction storms come from might make some places wetter and others drier. And since the country generally has no moisture surplus—Athens now averages about 15 inches of rain per year—a small decrease in precipitation any place in the country might be devastating.

But while Carpenter pointed out that the same amount of rain would not fall everywhere in Greece, and while he suggested that the known population pattern of 1200 B.C. might have been the result of a dry Mycenae, a well-watered Athens, and so forth, he could not show that rain had actually fallen according to his proposed pattern in 1200 B.C.—or ever.

Since this puzzle involves questions of climate, how might the science of climatology approach it, and what could climatologists say about drought in ancient Greece? If the techniques of climatology can shed some light on the decline of Mycenae, they might also tell what worldwide climates are possible in the future, and how those climates could affect mankind.

The questions for climatology to attempt to answer, then, are:

1. Is it possible to have the kind of drought pattern in Greece that Carpenter proposed?
2. What evidence do we have about what the actual conditions were when Mycenae declined?

Is it possible?

To answer this first question, climatologists at the University of Wisconsin-Madison who became interested in the Mycenae question had a choice of techniques. For instance, we could describe the topography of Greece to a computer, and with appropriate programming let imaginary winds blow across simulated mountains. We could try moving various storm systems across Greece from different directions, and see where the rain fell and did not fall.

That might be an interesting exercise, but it could not give proof. Behind any answer to our question, "What are the possible rainfall patterns in Greece?" would always be another question, "How good is the computer model?" How many lakes, rocky hillsides, and coastal inlets would we need to insert? The computer might print out a dozen patterns of actual rainfall that existed in Greece at some time or other, and one phony pattern. We wouldn't be sure which was which.

A better approach is to look at real patterns of past rainfall. This approach argues, simply, that what has happened, *can* happen.

To know whether the "Carpenter" rainfall pattern is possible in Greece, we tried to find whether it had ever existed. But we have detailed weather records in Greece only for this century. In fact, continuous records for more than a few locations there go back only to 1950, offering us a narrower range of possibilities than we would like.

We had 17 years of records (1950-66) for 17 stations in Greece. We took the "normals"—the averages—for that time, and compared each month to the 17-year average for that month. For instance, we asked whether January 1950 was a month of heavy, average, or light precipitation in Athens, as compared to the average of all 17 Januaries. We looked not just at particular places like Athens, but rather at how the pattern of regional rainfall changed from month to month.

Generally Greece receives most of its rain in the winter, so that is the crucial time for agriculture. For each winter month during the

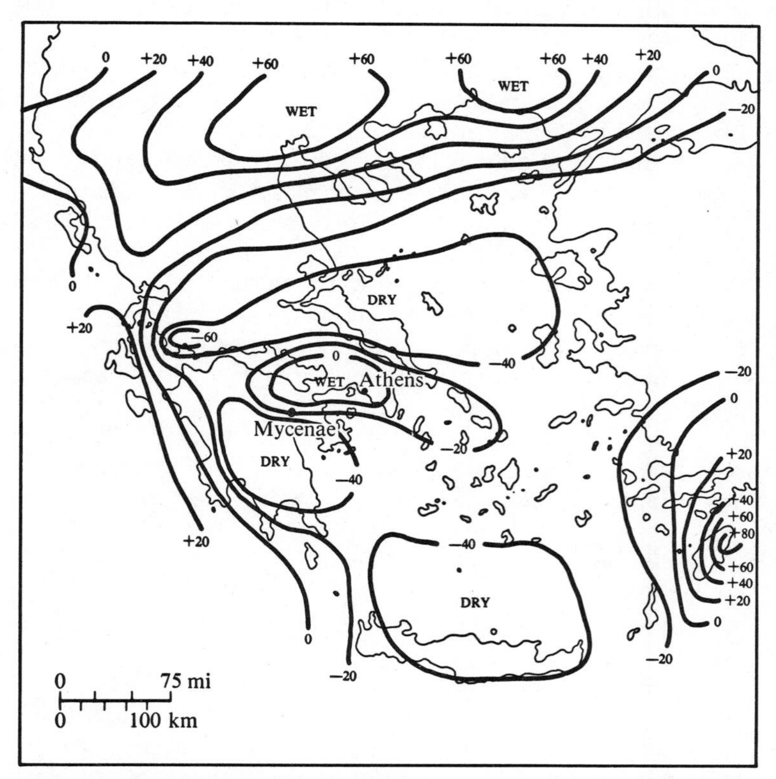

Figure 1.2. Rainfall in Greece and nearby areas: comparison of January 1955 rainfall with average of Januaries 1950-66. Numbers indicate percentage changes: -40 indicates rainfall in January 1955 40 percent below January 1950-66 average. After Bryson, Lamb, and Donley, 1974, p. 48.

17 years, we produced maps that compared rainfall in that particular month to the 17-year average for that month of the year (Donley, 1971). One such map, figure 1.2, the rainfall for January 1955, proved to fit the map of population shifts in Mycenaean times quite well. Mycenae and the Peloponnese were dry. Athens had above average rainfall. The island of Crete was dry, as was the region north of Athens. The islands of Corfu and Cephalonia, off the west coast of Greece, where population increased, had wet or "normal" conditions.

But what does the weather of one month, however well we know it, tell us about a drought which, if it did affect Mycenae, must have lasted for many years, perhaps many decades?

The rainfall pattern for January 1955 indeed was not the dominant pattern for all Januaries of 1950–66. It persisted only in that January and at some other times throughout that winter. During the 17-year span, three other winter patterns were more important. But although other patterns have been dominant in our own century, there is no reason that the 1955 pattern could not have dominated for years or decades in the past.

In fact, a number of patterns which are now uncommon have been common in recent centuries, as later chapters will show. Think of an especially dry August in your own experience, or a very snowy March. Suppose that *most* of the Augusts in your lifetime were that dry, or most of the Marches that snowy. Imagine that the conditions of the one "abnormal" case were in fact the "normal." That is the way we are using the January 1955 pattern. It is an example—an analog—of a climate pattern that may have been more important in the past.

Any pattern of rainfall in Greece, as we will explain shortly, is produced by a larger-scale pattern of winds that blow across the Mediterranean region, and that indeed are part of a global circulation pattern.

We had found that a pattern of weather does occur in which dry and wet areas are arranged in the way Carpenter had described. It is possible for Mycenae to be dry and Athens wet, and so forth. The pattern had moved from the realm of speculation to that of fact.

Did Mycenae decline in drought?

The answer to the second of our two questions must contain an element of speculation, because we simply don't have enough information from Mycenae itself. We are not, however, without a substructure of fact. Climatologists have some other tools to work with, in addition to looking for patterns in modern weather. Nature, as well as man, leaves records behind for us to decipher, and some of these reflect climate. The sediments deposited on lake bottoms millennia ago contain pollen of grasses, cattails, olive trees, whatever plants grew nearby. To discover layered, pollen-rich sediments is to discover a record of climate (though a less than perfect one).

Grains of pollen are immensely durable. Ancient pollen under a microscope reveals what plants produced it. Moreover, since vegetation responds to climate, prolonged drought will be reflected in pollen records. Olive groves, for instance, will eventually be replaced by plants that need less moisture than trees.

Pollen records are not available from Mycenae itself. But some records, dated to the time of Mycenae, are available from lake sediments in that part of the world (Wright, 1968). They show that the southern Dalmatian coast, about 250 miles northwest of Greece in what is now Yugoslavia, apparently had no change in vegetation. Nor did Pylos, about 100 miles southwest of Mycenae, on the coast, nor did Ioannina, in northwestern Greece.

Some reviewers of Carpenter's work, including Wright (1968), believed this showed that drought did not destroy Mycenae. But Carpenter had not claimed drought throughout the eastern Mediterranean.

Clearly, the records of past climates are difficult to read. Even a reliable record of climate in one place does not necessarily describe the climate 50 miles away. A drought for one city may accompany increased rain for others, and no change at all for still others.

The finding that rainfall was unchanged in a certain spot, Pylos, or Ioannina, or the Dalmatian coast, may be evidence *for* a particular set of changes elsewhere. Our map indicates that *no* change in these places might accompany drought in Mycenae. (The map also suggests that the agricultural area to the east of Pylos might have been dry, which would have taken a toll in that Mycenaean city.) The pollen records from these places, therefore, did not disprove Carpenter's theory of drought in Mycenae, and in light of what has been observed about rainfall patterns in Greece, the records could be regarded as evidence *for* the theory. But there is still more evidence available that bears on the question of drought as the cause of decline in Mycenae. That evidence comes into view if we step back and take a more global view of climate.

As part of the broader picture, consider winter storm tracks across the eastern Mediterranean—the paths of the storms that bring rain to Greece.

We compared the storm tracks for January 1955 with the average for 1950-66. In most years, storms traveled right through the region of Mycenae. But in 1955, the average track was 100 miles or so to the north (see figure 1.3). Since the rugged terrain of Greece causes a spotty rainfall whatever path the storms take, there was not a simple, direct, 100-mile-northward shift of all rains. What did emerge was the drought pattern we saw in our rainfall map.

A storm-track shift like the one of 1955 is not extremely obvious. To take an American example: a storm track which once passed over Chicago shifts northward, so that it generally passes over Green Bay, Wisconsin. Not every storm will follow this track, of course, but many will follow it or one close to it. Most people would not be aware that the shift had occurred, even if they watched weather maps. But the result might be a decrease in Chicago's precipitation—perhaps from two and a half inches in a given month to two inches, or an inch. This new pattern might dominate for only a few weeks, or it could come to be the typical pattern of whole decades. If it only lasted a month or so, it would receive little attention. If it lasted decades it would profoundly change a vast area of our Midwest. Such, we contend, was the case in Mycenaean Greece. But for more evidence, we must turn to other regions of the world.

Westerlies: links around the world

The storm track through Greece is not an isolated piece of weather set apart from the rest of the earth. It is one feature of the atmosphere's general circulation. The atmosphere observes no

Figure 1.3. Storm tracks in the Mediterranean, January 1955 and average of Januaries 1950–66. The January 1950–66 track, passing over the rugged terrain, brought more rain to Mycenae than did the January 1955 track. After Donley, 1971, p. 63.

national or regional boundaries, no beginning and end points. Its intricacies are many, but they are all related. A particular position of the Mediterranean track must link up with a specific set of weather features elsewhere. And a shift of that storm track must involve a shift of weather patterns in many places around the globe.

Climates are most directly related through a continuing westerly air flow, at high altitudes, around the earth's poles. The importance of

Figure 1.4. A polar view showing one possible configuration of the westerlies. The band indicates their outward (southernmost) edge. The pattern is not always the same, but whatever the number of loops and their positions, one flow links the hemisphere together climatically.

these westerlies and the nature of their influence will become more apparent as this book progresses. Since we are considering climates of the Northern Hemisphere, we will look at the flow of westerlies around the North Pole (see figure 1.4). The flow around the South Pole is roughly a mirror image. The details of this high-altitude flow of westerlies will be discussed later. For now, consider that near their outer edge lies the storm track we have talked about. The jet stream is also near this southern edge of the upper air flow.

All climates do not change in the same direction at the same time. The westerlies do not necessarily bring a cold winter to England and Iowa simultaneously, or cause a wet spring in Alberta to coincide with one in central India. Even in the limited area of Greece, this holds true. But as we also saw in Greece, weather takes on certain patterns, and so does the general circulation of the atmosphere. If January 1955 in Greece resembled a longer period about 1200 B.C., January 1955 should have resembled 1200 B.C. all around the hemisphere, not only in Greece. Such a deduction opened to us many more opportunities to test a theory about drought in Mycenae, where climate data are lacking. In testing our idea, we were no longer restricted to Greece alone.

Furthermore, since pollen and other ancient records show the weather of decades or centuries–they are not detailed enough to show individual months–any 1200 B.C. pattern they could reveal today must have lasted long enough to affect the civilizations of that time.

Comparing January 1955 to 1200 B.C., worldwide

One region for which we had some information for both January 1955 and 1200 B.C. lies across the Aegean Sea, in what is now Turkey. Asia Minor was drier than normal in January 1955–it received between 20 and 40 percent less rain than normal, depending on the area (Donley, 1971). Was it dry in 1200 B.C.?

Rhys Carpenter points out that the Hittite empire, which had ruled the Anatolian plateau from about the fourteenth century B.C. and was strong enough to have challenged Egyptian domination in the Middle East, declined quickly about 1200 B.C. Hittite mythology abounds in sun and storm gods; weather was always a chancy matter on the Anatolian plateau. And written records speak of famine, both of gods and men, near this time. Sometime near 1200 B.C., the Hittites sent an urgent request for food to Egypt, then a treaty partner, "lest we should starve." Relatively soon after that the Hittites appear to have abandoned the Anatolian plateau, moving to northern Syria (Carpenter, 1968, p. 46).

So we do have historical records of famine for the Hittites. And we have an archaeological record that strongly suggests drought caused the famine and the resulting population dispersal. There are also some pollen records available from Lakes Abant and Yenicaga in Anatolia; the pollen analyst reported that "the forest changes indicated by pollen analysis could . . . have been due to climate . . ." but that dating in the samples was imprecise (Donley, 1971, pp. 69–71).

If the proposed Mycenaean drought pattern existed, would it make sense for the Hittites to move to northern Syria? Again, we turn to winter 1955; in northwestern Syria and south central Turkey near Tarsus, precipitation was 40 percent above normal and temperatures were only 0.4°-1.2°C (0.7°-2.2°F) above normal. The Hittites would have praised their gods at finding such relief.

Thus we had another link between the winter climate about 1200 B.C. and the climate of January 1955, and the theory of drought in Mycenae was reinforced.

Examining the currently available climatic evidence of 3,000 years ago, worldwide, we did in fact emerge with the kind of correlations to the 1955 data that we were seeking. The evidence is, we think, the more convincing in that it was gathered by different scholars with varying interests (usually not climatic), yet it does form a consistent pattern. Here are some of the findings:

> Hungarian civilizations were disrupted by flooding about 1200 B.C. During the winter of 1955, the Hungarian plain received 5 to 15 percent more precipitation than normal. Perhaps more important, the nearby mountains had about 20 percent above normal winter precipitation and runoff from them produces spring floods.
>
> The Owens valley in California lies between the Sierra Nevada on the west and the White and Inyo mountains on the east. At the southern end of this valley is Owens Lake, now dry, but apparently filled to overflowing about 1200 B.C. In winter 1955, this region of California had above normal precipitation and below normal temperatures.
>
> In Norway, the snowline moved down the mountains in 1200 B.C. In 1955, Norway had below normal temperatures and normal or above precipitation. While summer conditions are also important for snowlines, actual observations from 1954 through mid-1955 were that the snowline did move down (Donley, 1971; Bryson, Lamb, and Donley, 1974).

We found no cases that were not consistent. In short, a drought could have caused the decline of Mycenae. We have proof that the proposed Mycenaean drought pattern can exist. We also have evidence that the pattern did dominate about 1200 B.C.

We may never know for sure what the climate was at Mycenae 3,000 years ago. But we can well believe the priest of ancient Egypt, even though the story he related was in mythic guise. We can also believe that climatic changes can destroy, or aid, one city or civilization. As our story moves to the present time, closer to certainty in our information, and closer to the lives of us all, we will see that humans over the centuries have been vulnerable to changes of climate, and remain vulnerable.

CHAPTER **2**

The Case of the Missing Farmers

STRETCHING ACROSS the Great Plains from Iowa to Colorado, drifted over with the soil of centuries, lie the vestiges of a thousand small villages. They are lost to the sight and mind of all but a few archaeologists, but once they teemed with life—the bustle of hunting preparations alternating with the care of fields and the tending of corn. For generations the cycle of seasons and the cycle of life went on, summer alternating with winter and cultivation with hunting. The debris of the hunt and of daily chores accumulated in and around the villages, to be smoothed out occasionally with a general housecleaning, gradually building layer on layer with remains of generations past—pots and hoes, bones, and occasional charred cobs and kernels of corn.

In the sixteenth century, when Coronado traversed the plains in his vain search for the Seven Cities of Cibola, he found no cities, and very few of the small agricultural villages that had once dotted the area.

In the early nineteenth century, when the mountain men and explorers who spearheaded the European invasion of the American heartland crossed the plains, they found no corn-farming villages. They left behind the last of the agricultural tribes as they moved out onto the grasslands—the Arikara and Mandan on the Missouri and the Pawnee in eastern Kansas—not to find corn fields again before reaching the Pueblos in the southern Rockies. They traded with nomadic tribes and hunters who had lived in the area for centuries.

In the twentieth century, remnants of the villages were uncovered, but only as layers of accumulated debris covered by windblown soil. Where had all the farmers gone? When had the villages been abandoned? And why?

The fire of St. Anthony

Several centuries before Coronado and before the mountain men, from the ninth to the fourteenth century, a strange madness accompanying a devastating illness periodically reached epidemic proportions in western Europe. Whole villages would suffer convulsions, hallucinations, gangrenous rotting of the extremities, and often death. Pregnant women would abort, and even pets and domestic animals would give evidence of the same illness and die. In its acute form the disease brought great abdominal pain and violent convulsions, followed by speedy death. In the chronic form it affected the extremities: an icy chill developed, followed by a burning sensation. The limbs darkened, shriveled, and fell from the body. The affliction was so widespread that monasteries were devoted to the care of the unfortunate sufferers who did not immediately die, and in 1096 the order of the Hospital Brothers of St. Anthony was founded to care for the victims of the disease. Because the blackened, gangrenous hands and feet looked as though they had been burned, the disease was likened to fire; it was often called St. Anthony's fire.

In 1596, though the incidence of the disease was greatly reduced, the medical faculty at Marburg identified the poison which had wrought such havoc (Haggard, 1929, p. 218). It was contained in kernels of rye affected with ergot blight—the fungus now known as *Claviceps purpurea*. The blighted kernels are blackened and enlarged, and only a few incorporated into a sack of flour are enough to affect those who eat the bread made therefrom.

St. Anthony's fire last appeared in 1951 in southern France, and investigation there finally solved its most puzzling feature—the hallucinations, which had, of course, been given a religious interpretation in earlier times. Blighted grain, if stored damp, ferments slightly and besides ergot produces a second drug—a form of the hallucinogenic drug now known as lysergic acid diethylamide (LSD).

Why should a fungus like *Claviceps purpurea* be particularly prevalent across Europe, especially in eastern France, from the ninth to the fourteenth century, then diminish? Ergot blight develops in cool, persistently damp weather. This suggests that such weather was common in western Europe between the ninth and fourteenth centuries.

There is documentary evidence of such weather as well. From manuscripts and chronicles it is possible to extract pieces of weather information which, carefully collated, yield an outline of the changing weather patterns of the last millennium. There are few references in medieval sources to very wet summers, but there does emerge a pattern of summers when fields were not dry enough to be worked or to permit grain to be harvested. About 1300, farmers abandoned many fields in England, Denmark, and other parts of western and northern Europe. Pollen records from the same time indicate a shift to a wetter climate (Steensberg, 1951).

The winters were generally mild until the latter part of the sixteenth century. To Europeans of today this mild, wet pattern is well known, and they rather uniformly detest it. In Germany it is called the *Westwetter*; it sometimes hangs on for what seems a gloomy eternity, and is the result of steady west winds from the Atlantic carrying moist air far into the interior.

In winter it means little snow cover in the Ukraine, with winterkill of wheat when the inevitable frosts do occur. Farther west, the vines and fruit trees often bud out and then die if the frosts come; one severe frost may do it. In summer the Westwetter produces lush growth, but little chance for the bright sun and drying weather that are also needed for a good grain harvest.

What does this have to do with the missing farmers of the plains?

The boreal forest

In North America, the boreal forest, the spruce forest of the north, covers a well-defined climatic region. The spruce forest must have arctic air in winter. In summer it must have Pacific air, dry air coming down from the western mountains. Farther north, under arctic air year round, lies tundra. To the south of the spruce, where moist tropical air reaches in summer, are forests of mixed hardwoods. Or, where dry air from the west prevails, lie grasslands.

The forest border, especially that part east of the Rockies, is a relatively smooth line. One reason for this is that one of the loops of the westerlies is generally "anchored" over the Rockies, which are high enough to have this effect on their flow. The airstream flows eastward from there over generally even terrain. Therefore the climatic border (and the forest border) is less complex than are the climatic patterns in many parts of the world–for instance, in mountainous Greece.

When the boundaries of arctic air change, the forest boundaries move too. The records of these shifts lie in the soils produced under forest and tundra. From the dropping of needles and cones and the physical and chemical actions unique to such a setting the forest produces a distinctive soil, called podzolic soil. North of the present forest, in a zone stretching 1,000 miles from west to east, are many areas where layers of podzolic forest soil alternate with the grey-brown soils of the treeless tundra.

In these soils are written the comings and goings of the forest as the arctic air environment expands and contracts. A thousand years ago the forest had advanced northward into the barren lands, the tundra. Then something happened to reverse this advance. The Arctic was expanding again, and for the area that lies generally north of the Great Plains, this last major expansion is recorded in the soils from about the beginning of the thirteenth century (Bryson, Irving, and Larsen, 1965; Bryson, 1966).

Detectives and scientists

In the early 1960s, these seemingly disconnected facts began to raise a whole new set of questions. Why did the farmers disappear? Were the villages emptied at the same time St. Anthony's fire mutilated and killed Europeans? Could the soils of Canada give a clue? Climates in some parts of the world appeared to have changed about the time the farmers disappeared—although some of the dates were not yet firmly known. Perhaps a changing climate was the farmers' downfall.

At that time no one had asked these questions because no one had considered the fire, the villages, and past climates together. St. Anthony's fire is a historical footnote; the plains Indians of the eleventh to the fourteenth centuries are hardly even that outside a small group of archaeologists and regional history enthusiasts.

But the case of the missing farmers and the clues—or potential clues—that appeared from time to time were intriguing to a climatologist.

The analytic approach of the scientist, it has often been remarked, resembles the modus operandi of the detective. Both work by formulating a hypothesis to explain some observed facts (how the crime, or the event, *might have* happened) and by refining that theory as further facts accumulate (how the evidence shows it *did* happen). But the scientist—in climatology or any other field—must take the process one stage further. For reliable conclusions, he must *predict* events, then conduct an experiment to test his prediction, to show how well he has understood the workings of nature or man. The prediction can be a "forecast" of the past—if the events are not yet known but can be learned. Moreover, it must be possible to repeat the test, or to run additional tests, for confirmation. If confirmation fails, the scientist must go back to the beginning.

The prediction-test-conclusion feature of the scientific method is important. A chemist, for instance, thinks he sees a principle at work. He predicts how it would apply in a certain case, and runs an experiment. He combines certain chemicals and sees whether the resulting compounds are the ones he predicted. But he must make a real prediction. If he merely "predicts" a chemical reaction he has seen before, he is not testing his hypotheses about the cause of certain reactions.

Not all investigations are conducted this way, of course. Another approach is to attempt to explain or rationalize events without a prediction. In effect, this approach is to run the experiment first, to "see what would happen if . . . ." Then the findings are explained, or at least rationalized. When the chemist or the climatologist operates this way, however, the pitfall is that some explanation can always be devised to more or less fit the results. Until the explanation has met the test of repeatedly predicting events, it is unproven.

The a priori approach, reasoning from cause to effect, had to be part of the detective work on the missing farmers, if that case was to build knowledge and theories that would be generally useful.

On the case

A colleague, David Baerreis, a University of Wisconsin anthropologist, also was interested in the missing farmers, and we began to investigate. We did so several years before the study of Mycenae, so this case broke new ground. And, as can be expected with new ideas, we had surprises coming. The path toward our conclusion could not be foreseen. What we found, however, about the climate of 800 years ago was as revealing as what we found about Mycenae and the world of 2,400 years before that.

Our aim was to test ideas about climate in general, and to solve the mystery of the missing farmers. We decided to proceed by examining data worldwide and trying to discern a particular atmospheric pattern.

Then we would test this hypothesized pattern. We would (1) predict climate in the Great Plains for some specific time around A.D. 1000-1200; and (2) excavate for climatic evidence at some site where the missing farmers had lived.

To begin, we needed more precise details of European climate in order to establish a pattern for the westerlies. Tales of St. Anthony's fire were hardly specific, and weather-instrument data no more existed for this time than they had for Mycenae.

But medieval Europe had left some records of climate, almost inadvertently: legends, national and family chronicles, vintners' records of the grape harvest, church records of prayers for rain (Lamb, 1966; Ladurie, 1971, pp. 270-287, 379-386). Any product of the past, including poetry and paintings, is a potential source of information about climate.

These records do have limitations; few if any were intended to describe climate. But while any one record or set may be suspect, a meticulous compilation of everything available should be taken seriously if it points in a particular direction. We had the information we mentioned a few pages back: fields too wet for grain harvest, abandonment of fields in England and Denmark about 1300. More detailed information exists: some of the most useful comes from two British climatologists, H. H. Lamb and Gordon Manley. Lamb's and Manley's work emphasizes historical records. But along with written references to the weather, the historian of climate can also turn to pollen records, tree growth as shown by rings, and all the other traces of past climates that are available. Manley (1953, 1974) put early thermometer records into useful form.

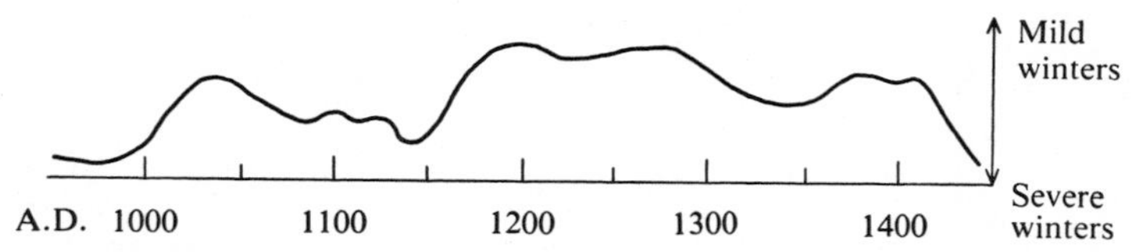

Figure 2.1. Mild and severe winters in western Europe, A.D. 950–1450. After Bryson and Baerreis, 1968, p. 33; based on Lamb, 1966.

Lamb (1966) brought together his evidence to show the complex pattern of climates for many places over a span of several centuries. We will look again at some of his work in a later chapter. For now, we are concerned with his findings about winter mildness in western Europe over roughly the first four centuries of this millennium, a time that, we believed, included the abandonment of the plains villages in North America. The findings form the basis of figure 2.1. This reading on European climates doesn't attempt to define conditions very precisely, but it does show changes during the centuries of interest to us. We would, then, attempt to match changes in the severity of European winters, as shown by Lamb, with some climate change in the North American Great Plains.

What change, and when?

But the time span of interest was several hundred years. In what particular time did we want to focus? For what time, if any, would we find environmental problems for the plains Indians? We used two clues within our information about Europe and about the boreal forest in Canada to focus our prediction in time. In both places, the 100 years or so around 1200 saw significant changes in climate: in Canada the forest moved south; the notable feature in Europe's cli-

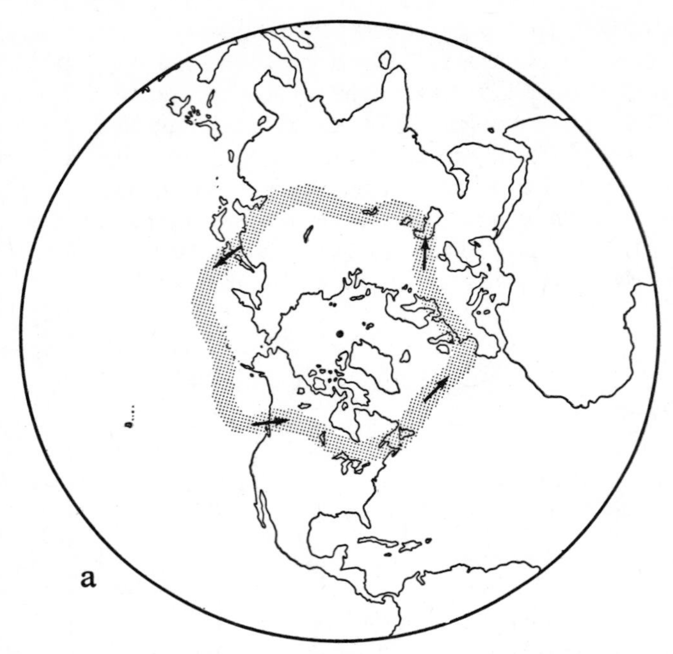

Figure 2.2. Patterns of the westerlies. They can form a rather tight circle around the pole, typical of summer (a); they can follow a much more southward and looping course, a winterlike pattern (b); or they can take other routes. The shaded area is the southernmost part of the westerlies, the region where they are strongest.

mate, according to Lamb's data, was a marked shift toward mild winters, beginning about 1150 and lasting a couple of centuries or more.

At this point, we began to see that the outbreak of St. Anthony's fire was not an essential clue in our case. The disease, in fact, appeared from time to time over many centuries. In Europe, references to it go back to well before A.D. 1000 and it did not pinpoint the major climatic change we thought it might. A false clue, in a sense, and yet one without which we might never have examined the case of the missing farmers.

We now had more substantial evidence for climatic change than the "Fire." The question was: what general circulation change around the hemisphere—what particular pattern of the westerlies—would cause the winter mildness about 1150 or 1200?

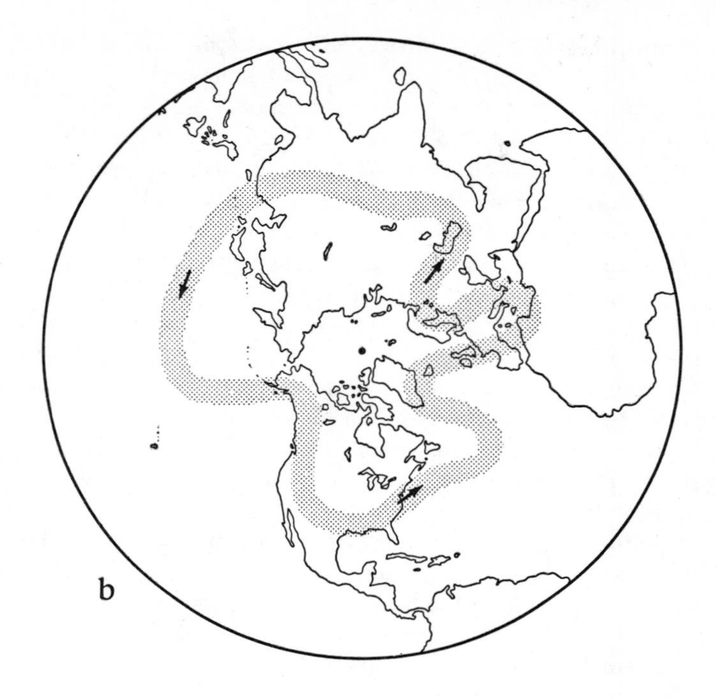

Westerlies, wet and dry

The westerlies, which contain the jet stream, and are part of the general circulation of the atmosphere (all those names are used), consist basically of a flow of air circling the earth's poles–a flow strongest at an altitude of 30,000 to 35,000 feet.

As the westerlies sweep around the hemisphere, they do not take a smooth, even, west-to-east path, but rather make north-south waves, or loops, in their course. And they do not always flow at the same latitude. Seasonal changes are apparent: in winter the westerlies generally have their most southward position. Arctic air dominates a large region. In summer the westerlies as a rule are more contracted; polar air (not as cold as it is in winter) is usually confined to the far north, and warm tropical air, which can push up to the edge of the westerlies, dominates the temperate regions (see figure 2.2).

The pattern of loops in the westerlies varies also. The number and position of the loops, and the extent of their north-south swing, change from time to time. When the westerlies are expanded, these loops are more likely to be great sweeps, so exaggerated that northerly and southerly flows of air are brought to many regions. But the general circulation in such cases, if one sees its course on a global scale, is still westerly.

The westerlies do not change in lockstep with the seasons. The looping flow can take place in summer or in winter. And a pattern once rarely seen in winter can come to be the typical winter pattern for decades or even for centuries. As we pointed out in chapter 1, a pattern of westerlies that steers winter storms over Mycenae may be replaced by a pattern that takes the winter rains farther north.

So the westerlies' influence on climate is complex. At a given time, the place where you live can be within the circle of westerlies or outside it. It can be under a flow of high-altitude air from the west, north, or south, depending on the arrangement of waves. Local or regional patterns can produce areas of easterly flow. In our attempt to understand European and North American climates for A.D. 1200, however, we tried to reduce the complexities as much as possible, to deal with the essential elements.

The source of damp, drizzly weather in western Europe is the North Atlantic. Winds sweeping in from the west have picked up moisture from the ocean. They are neither coldly arctic nor warmly tropical. A concentration of the westerlies at the latitudes of Europe, about 40°-50° north, brings the gloomy summer days that keep farmers out of the fields, and the damp mild winters that bring forth buds too early. It brings the Westwetter, and encourages St. Anthony's fire. More precisely, the Westwetter comes when the westerlies are slightly expanded—farther south than at other times—but not greatly expanded, with strong north and south loops. When the Westwetter leaves, it is because the westerlies are farther north, or are looping so that Europe is not receiving its weather almost entirely from the Atlantic.

This pattern of circulation became our hypothesis.

The soils of the boreal forest in Canada reinforced this hypothesis. An expansion of the pattern of westerlies, the flow around the pole, would bring arctic air farther south–and cause the forest to retreat from its northern boundaries, as it did about A.D. 1200.

Suppose the westerlies did expand slightly, pushing the boreal forest south and bringing stronger west winds to the latitude of Europe. What would happen to the villages of the North American plains? We believed the slightly expanded westerlies would bring a flow of air sweeping down from the Rockies and across the plains. And, as is always the case with air coming down a mountain slope, that air would be dry. The conditions would be the same as those Rhys Carpenter noted for Greece: as air descends from a mountain range, the relative humidity decreases.

Since westerlies are a continuing feature of the atmosphere, we find a continuing dry "shadow" east of the Rockies–the Great Plains. But when the westerlies are generally contracted toward the pole, moist air can penetrate northward from the Gulf of Mexico and the shadow is relatively small.

In A.D. 1200, we reasoned, the expanded westerlies would have pushed eastward, blocking off sources of moist air, enlarging and intensifying the dry shadow of the Rockies.

While western Europe grew damp and gloomy, the farmers of the plains 800 years ago must have seen their corn wither and turn white with drought, their game die or move away.

We had our hypothesis and our prediction; we still had to put them to the test.

CHAPTER 3

Weather of Indian Times

WHEN WE PROPOSED a drought for the Great Plains 800 years ago, we made, in effect, a prediction, though one for the past rather than for the future. But the prediction was not specific enough to be tested. It had to be refined: where, precisely, would slightly expanded westerlies cause a drought, and how serious would it be?

The weather map for A.D. 1200

We used the same technique described in chapter 1—that of past climate analogs. We found modern times of expanded westerlies and checked the resulting rainfall patterns.

While such expanded westerlies have not dominated decades in modern time, they have governed the weather of individual months. We compared the weather data from those months to data from months with a more contracted pattern, and mapped our findings (figure 3.1). Such a map shows the details of a dry *month* in modern times, as a model of a dry decade or century in the past. As we

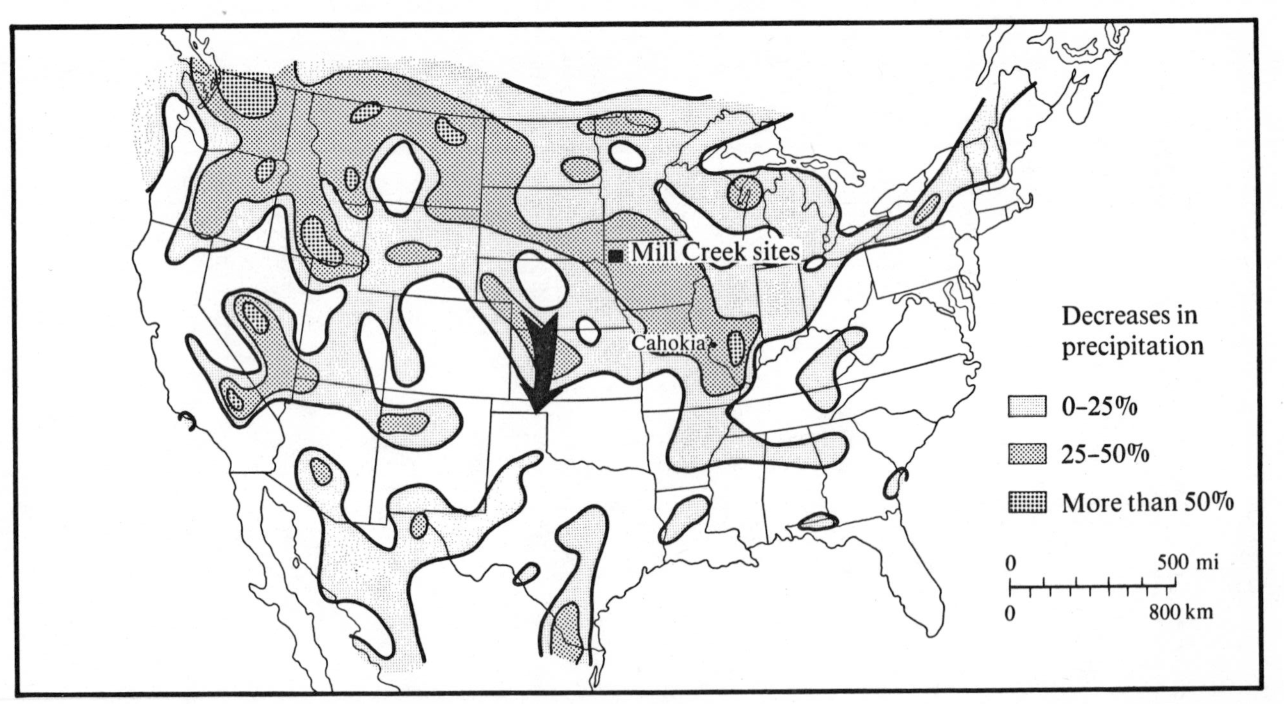

Figure 3.1. July precipitation decreases to be expected with a slightly expanded flow of westerlies, based on 20 years of modern weather records. Shaded areas have less precipitation when westerlies are expanded. Unshaded areas have various percentages of increased precipitation at the same time. Arrow indicates migration of Upper Republican Indians, discussed later in this chapter. Adapted from Bryson and Baerreis, 1968, p. 7.

pointed out in the case of Mycenae, such an analogy can be drawn because the same atmospheric pattern produces both times of dryness; the difference is in how long the pattern persists.

We focused on conditions in July, which is a crucial month for plains farmers; dry Julys damage agriculture there, just as dry Januaries are crucial in Greece.

Our map's most interesting feature was an area already known to students of the plains' ecology and climate, the prairie peninsula. This finger of dry climate and corresponding vegetation extends eastward from the Dakotas through Iowa and into Illinois (Borchert, 1950). This is the dry shadow of the Rockies, and it lengthens and becomes drier as west winds blow more strongly.

Our map showed that in times of increased westerlies, the prairie peninsula received little precipitation. Rainfall there was 25 percent or more below normal. We had, then, a prediction of drought in a specific area. To test the hypothesis, Baerreis and Bryson determined to excavate villages within this peninsula for evidence of climatic change.

As an anthropologist, David Baerreis brought to this work a specialized knowledge that was essential. Our cooperation points up an important trait of research on past climates: it must be interdisciplinary. The climatologist cannot afford to overlook any possible clues or techniques simply because they are outside his topics of work in the past.

An ecotone in Iowa

Usually, archaeology does not employ the scientific method of hypothesis, prediction, and testing. Archaeologists excavate promising sites, then analyze and interpret what they find in light of other evidence. The full scientific method was a fairly unusual approach to such a case.

We knew several sites of ghost villages within the prairie peninsula, including those of the Mill Creek culture in northwestern Iowa, first excavated in the 1930s, but not thoroughly explored. We knew that the villages had been occupied about A.D. 1200 by a hunting and farming society; we could hope for a variety of climatic signals among the thick piles of debris left behind (see figure 3.2).

That area today averages about 25 inches of precipitation a year. Generally it produces good corn and soybean yields, but the rainfall

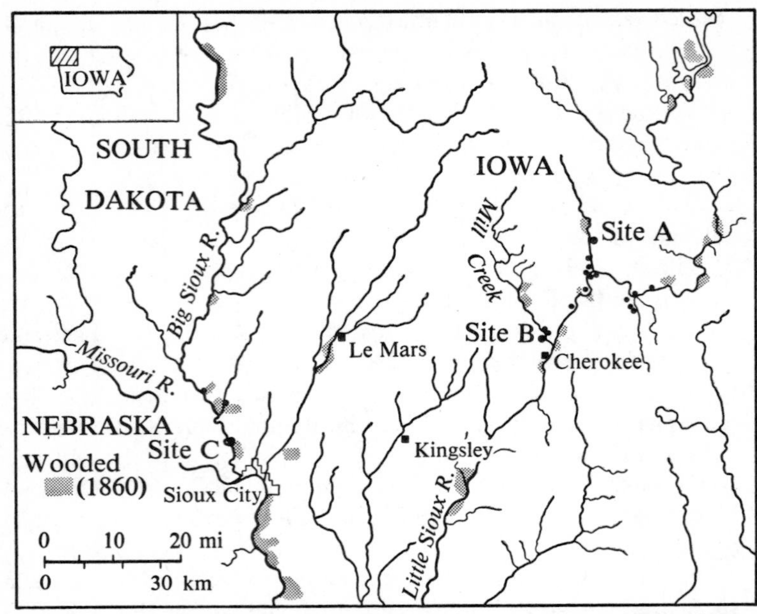

Figure 3.2. Area occupied by the Mill Creek people about A.D. 900–1400. A, B, and C were the principal research sites discussed; unlabeled dots indicate other known sites. After Bryson and Baerreis, 1968, p. 291.

is not always enough to compensate for hot summer winds, and agriculture is frequently moisture-limited. A drop of 25 percent or more in moisture will show up dramatically in cornfields; it would also show up in the hunting and agricultural economy of the Mill Creek people. Refuse of their diets would reflect a drought–as would the pollen scattered by plants growing 800 years ago.

Northwestern Iowa is a borderland: specifically an ecotone, or transition zone between two ecological communities. In this case one community was the steppelike short-grass region to the west, a more arid area. To the east more rain comes, supporting tall-grass prairie (and in modern times much of the tall grass called corn).

An ecotone contains a mixture of organisms from both sides. It is an area sensitive to change, a good spot to check a pulse–in this case, of climate. As our rainfall maps of Greece showed, you have to know where to look for evidence of climatic change. And *change* is what interested us. Indians had farmed and hunted in northwestern

Iowa, then disappeared. A constant climate, however dry, would not explain that.

Given our forecast of drought at Mill Creek, just what did we look for, and specifically what predictions did we make about our findings?

Vegetation

The plant populations of the ecotone area of Iowa are especially sensitive. More rain favors the tall prairie; less rain gives an advantage to shorter plants and those with less leaf surface.

We also wanted to consider two types of trees in northwestern Iowa. One group, which includes cottonwoods, often grows with roots in the water table and so can survive even prolonged droughts. Others, like oaks, grow in better-drained areas and need regular rains. All trees need a lot of water; a decrease in rainfall favors the first type, but also means fewer trees overall.

We predicted, then, that plants which could survive with little moisture became more dominant about 1100–1200. Our main source of evidence would be pollen records.

Animal life

Animal populations also reflect climate, because of their own climatic preferences and because of the vegetation needed in their food chains. Rather small differences can be important. We know that both bison and deer roamed the Mill Creek area. Bison eat grass and deer browse on trees. Trees should be hurt most by drought, and that would cut the food supply for deer.

The deer population declined sooner and to a greater extent than bison herds, we predicted.

Human life and culture

The people who lived around Mill Creek depended on the plants and animals. A drought should affect them, just as serious drought brings suffering today.

Those Indians grew corn, which demands a good deal of moisture. We predicted problems in corn production.

The Indians also ate bison and deer, along with other game. Whatever their preferences, in hard times they would eat what they could get. In dry times bison should be easier to get than deer. The trash the people left behind, then, containing bones from their game, should be a way to gauge animal populations and would show human activity as well.

A Mill Creek calendar

We had predicted a drought about A.D. 1200. But putting dates on the remains of old cultures, especially nonliterate ones, can be as difficult as determining what happened there. For many civilizations, pottery sequences and levels of occupation can provide reasonably accurate comparative dating, but for absolute dates these sequences must be linked to objects that come from a time or a civilization that can be precisely dated.

Tree-ring counts can provide very accurate dates. It is sometimes possible to count back, one year per ring, beyond the age of any tree now living, by finding remains of long-dead trees and matching up the ring patterns of successively older trees.

But at Mill Creek we had none of these aids.

In constructing a Mill Creek calendar, we relied on radiocarbon dating, which takes advantage of the fact that a radioactive form of carbon atom in the atmosphere finds its way into living matter, along with much greater quantities of ordinary carbon, and decays at a known rate. Of a given quantity of radiocarbon, one-half will decay in about 5,700 years.

But radiocarbon dating cannot be precise. Actual measurement of radiocarbon content in a sample is difficult, and the radiocarbon content of the atmosphere has not been constant. Although Wisconsin climatology students constructed a chart of correction factors that makes the dating more accurate (Wendland and Donley, 1971), even our best readings in the radiocarbon tests for Mill Creek only tell what century a sample came from, rather than what year or decade.

We were, however, able to take many such readings for Mill Creek, and as the readings began to cluster around certain dates, we became more sure of those dates than we could be witn only one or two samples. This information, combined with that provided by the sequence of findings (new material goes on top of old, generally speaking) enabled us to establish a reasonably certain range of radiocarbon dates from about A.D. 900-1400 (Bryson and Baerreis, 1968).

The Mill Creek weather station

At the Mill Creek sites, a thin overburden (usually just a few inches) overlies material left by the missing farmers. Those deposits may extend down for a couple of feet, or as much as eight or ten feet. Digging in the villages of 800 or 1,000 years ago, our objective was often the town dump–bones from the hunt, broken pottery, litter from housecleaning, ashes from the campfire, and other refuse were piled together, or simply allowed to accumulate in place, along with leaves, dust, and (fortunately) pollen that drifted in.

Counting bones

The people of Mill Creek ate many kinds of game; at one site we found 10,675 bones, among them 229 from bison, 572 from deer, 13 from elk, 4,785 from fishes, 1836 from birds, 170 from rodents, 114 from carnivores, 43 from turtles, 34 from rabbits, and many unidentifiable fragments (Bryson and Baerreis, 1968, p. 300). We had enough deer and bison bones to focus on the relative proportions of those two. The tree-browsing deer thrive when good rains favor the trees; the bison depend on grasslands, which survive drier times.

To deduce the total importance of these animals in Indian diets, a scientist would have to consider more than just the number of bones. For instance, a bison weighs much more; its skeleton represents about five and a half times as much meat as a deer skeleton does. Also, complete skeletons are not found; pieces are. The scientist would have to decide just how many bones should be counted as evidence of one whole animal. We were able to pass over some of these details, because we sought indications of *change* in deer/bison proportions. We simply had to be consistent in our counting methods.

The diagrams that follow show some of the bone information we gathered. These figures are from excavation site B, with both deer and bison at all levels, and where each species dominated for a time.

As figure 3.3 shows, up to A.D. 1100 or so, deer bones ranked first at this site. After that, the percentage of deer dropped, and bison bones became more prominent–a shift that fits our predictions about climatic change.

Figure 3.4 shows that the total number of bones declined sharply

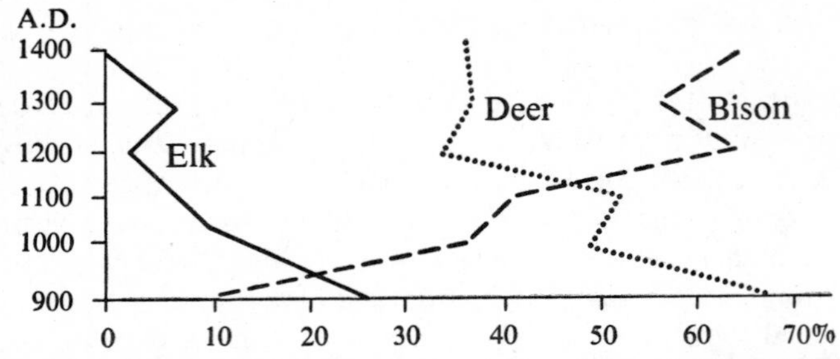

Figure 3.3. Percentages of elk, deer, and bison bones found at Mill Creek site B. After Bryson and Baerreis, 1968, p. 291.

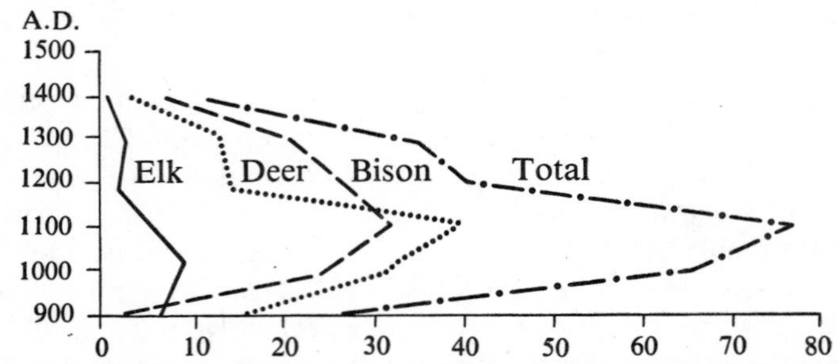

Figure 3.4. Number of animal bones found at Mill Creek site B. After Bryson and Baerreis, 1968, p. 291.

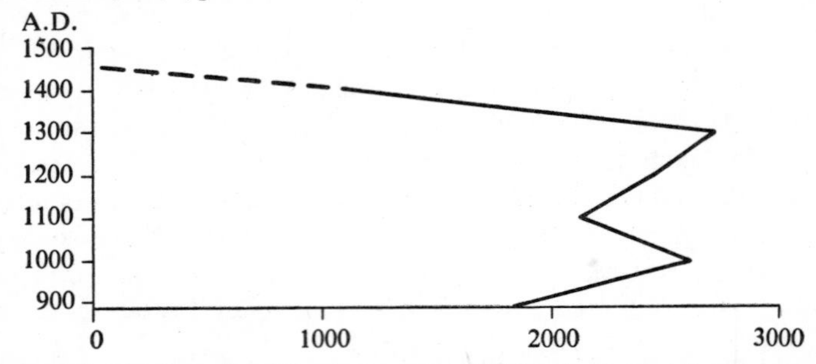

Figure 3.5. Number of potsherds found at Mill Creek site B. After Bryson and Baerreis, 1968, pp. 197–198, 290.

after 1100. A decline in deer and bison herds—fewer animals to hunt—might be one cause of that.

If drier weather came to the Mill Creek region in the twelfth century, the effect on trees would not have been immediate. But eventually some trees would die and fewer take their places; the trees remaining would not have lush growth. The prairie grasses would survive but they would not be so tall or so thick as before. So bison would survive better than deer, though both herds would shrink somewhat. In the long run, the Indians would have found hunting more difficult.

The Indians were corn farmers as well as hunters, and a drought should have affected their corn-growing too. As an index to corn production, we counted potsherds–pieces of broken pottery (Bryson and Baerreis, 1968, p. 22). We believed corn production and potsherds were related, because pottery would be needed to store, cook, and serve corn. Meat could be roasted and held in the hand.

The number of potsherds remained high when bone counts first dropped (figure 3.5). This hints that, while the drought must have hurt the corn farming, it hurt hunting even more. The potsherd count, if read as a reflection of the total number of people in the village, also suggests that the human population did not decline immediately.

But with little food of any kind as the drought continued, numbers of bones and sherds both dropped rapidly after 1200, and by 1400 or so there were none, and no Indians either. The farmers were gone.

Other villages

Other Mill Creek sites we studied, while not providing as great a range of habitat, also contained indications of climate past. Figure 3.6 gives a sense of three Mill Creek culture sites we examined in detail. We have been discussing the intermediate site, site B. The upper site, near the headwaters of a creek, had few trees and never many deer. It revealed the same general pattern, however: a decline in bones followed by a decline in potsherds. Our third Mill Creek culture site, labeled C, in the broad valley of a river, had deer and bison throughout the Indian occupation, though always more deer. It contained a great number of bones and potsherds; it was the most populous of the three. Village C seemed to peak about 1300, somewhat later than the others; the village held out longer. The more reliable water supply might have brought an increase in game as animals moved from drier places.

The peak of Indian occupation there did not last long, whatever its exact time. Between 1300 and 1400, according to our dates, the bone refuse dropped by about 80 percent.

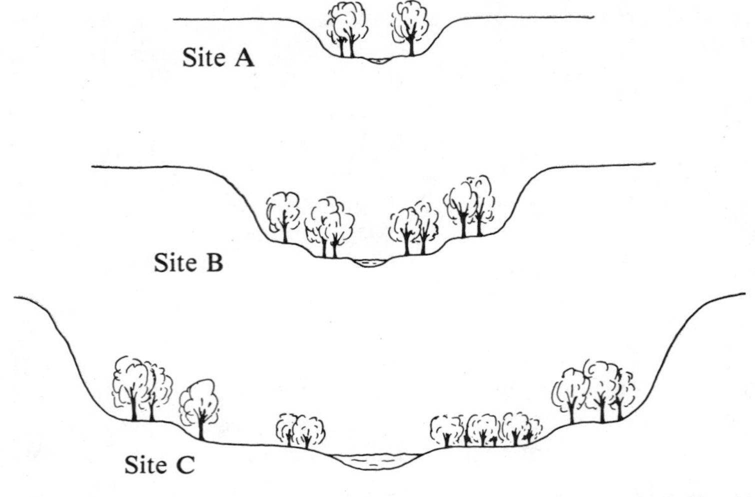

Figure 3.6. Schematic comparison of Mill Creek culture sites A, B, C. Site A: Few woods; much more attractive for bison than for deer; creek would dry up quickly in drought. Site B: Most interesting research site; attractive for deer and bison; good water supply during moderate drought. Site C: Heavy woods favorable for deer; water supply would remain even in severe drought; largest village.

The potsherds are also most numerous at the A.D. 1300 level, which we believe was well after the drought started. In fact, they are so numerous as to suggest another use, besides storing and cooking vegetables. Were they a means of irrigation? Along this river, were there a few years as the drought worsened when carrying water made a difference?

These three Mill Creek culture sites point up again the importance of careful interpretation in reconstructing past climate. We saw that pollen records from some parts of Greece show no climatic change for 1200 B.C., although climate changed nearby. The Mill Creek sites were not very far apart; they must have experienced the same progression of climates. But they each *reacted* differently to a climatic change, and their records must be read accordingly.

The prairie and the trees

As bones, broken pottery, and miscellaneous trash went into the pits and refuse piles of the Mill Creek Indians, pollen drifted in from prairie grasses and flowers and from tree blooms, and settled to the ground and into the trash pits.

Pollen layers in a trash pit are not so neat as in lake sediments. Trash does not accumulate in level layers, and some mixing takes place. Still, pollen records from Mill Creek may be more significant than bone and potsherd counts, because pollen cannot be changed by housekeeping habits or dietary tastes. Pollen describes the vegetation, and the vegetation tells us about climate.

Northwestern Iowa was mostly tall-grass prairie, though with some short grasses characteristic of drier, steppelike regions. There was Indian grass, six feet tall and topped by feathery golden seedheads; big bluestem, just as tall, with a three-toed "turkey foot" head; little bluestem, its stems red in fall and winter with fluffy tufts of seeds; needle grass, with seed carried by corkscrewed darts; and pasque flowers, shooting stars, sunflowers, cone-flowers, asters, and other blooms in the months from April to October. The midwestern prairies were highly developed communities of plants and animals, each adapted to take advantage of some feature of the environment. For instance, the prairie's succession of blooms throughout the season gives each plant an opportunity for pollination by insects and wind.

In the creek bottoms and on moist hillsides, trees also grew. In the bottomlands, where their roots often reached to the water table, were willows, elms, silver maples, cottonwoods, walnuts, and a few others. On better drained but still moist slopes grew red and white oaks, basswoods, sugar maples, and again, elms.

The best pollen records of these plants comes from the same site that gave the best bone data, site B (see figure 3.7). From the pollen contributed by trees, we found a dominance of the oak group at older levels (perhaps A.D. 900-1200). Above that, the willow group came to dominate. The prairie pollens from the same site changed at about the same time. They moved from a high percentage of composites–which include sunflowers and asters–to a larger percentage of grasses, which have smaller leaf surfaces and require less moisture; they are favored by a drier climate. In both cases a rise in the curve, a rise such as appears at about the middle of the period, is what we predicted for a time of drought. The sequence reflected in these two curves has an unmistakable resemblance to the graph of winter mildness in Europe (figure 2.1).

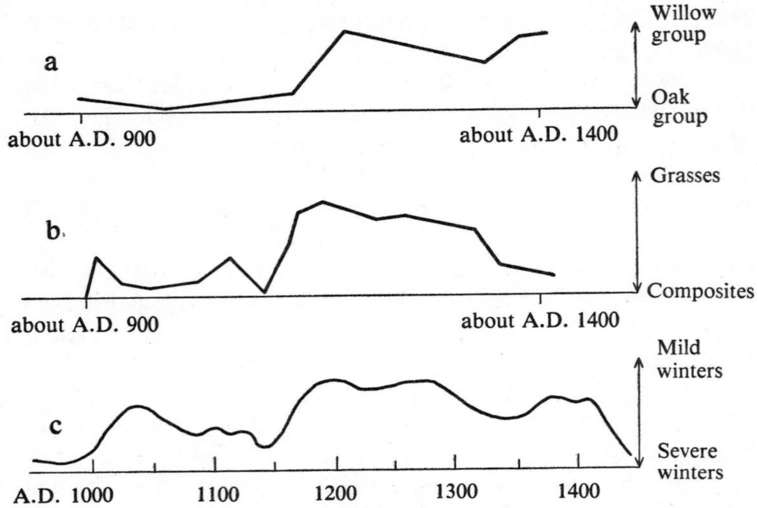

Figure 3.7. Pollen records from Mill Creek site B: (a) trees, (b) prairie pollens; compared with (c) mild and severe winters in Western Europe, A.D. 950–1450 (as shown in figure 2.1). After Bryson and Baerreis, 1968, p. 33.

Our prediction of drought in Mill Creek was supported. Our hypothesis of expanded westerlies had met an important test, and we had taken a step in demonstrating how the world's climates are linked and how climate changes.

Republicans in the panhandle

To test our views further, we looked for opportunities elsewhere on the Great Plains.

About A.D. 1000 corn farmers worked all across the high plains west of Mill Creek, from the base of the Rockies through eastern Colorado and western Nebraska. One group of these ancient farmers is called the Upper Republicans–they lived on what is now the Upper Republican River in western Nebraska for several centuries before A.D. 1200. This area is even more marginal, more quickly hit by drought, than western Iowa, and other researchers, we knew, had hypothesized that a thirteenth-century "dustbowl" had finished off those villages (Bryson, Baerreis, and Wendland, 1970).

This brought up an interesting question, because traces of the

Republican culture can be found in the "panhandle" region of Oklahoma. The remains of corn-farming villages there have Republican characteristics. Did these die out at the same time as those in Nebraska? The villages had not been dated; the answer was not known.

Our map predicting rainfall patterns for 1200 (figure 3.1) showed that while the Upper Republican River region should indeed have been dry then, the panhandle would have been wet. In that case, the best explanation was that, rather than stay and die in Nebraska, the Indians moved to greener fields. So we predicted that the panhandle villages did not die out about 1200, when the ones in Nebraska did. Rather, they should have *appeared* about that time.

Radiocarbon dates from three dozen samples in the Panhandle showed that the people did come about A.D. 1200 (Baerreis and Bryson, 1966, pp. 110–111). And they did not drift in slowly, over centuries. Our dates, and those of archaeologists who looked at trade pottery from more precisely dated cultures in the Southwest, showed that the immigration took place over 20 to 40 years, perhaps fewer than 20.

Monks Mound and the entrenchment of authority

In southern Illinois, just east of St. Louis, stood one of the major population centers of the world in A.D. 1000. The Indians who lived there grew corn in the fertile Mississippi bottom lands. Within a 125-square-mile area that included many settlements was the one now called Cahokia, with a population of perhaps 40,000. Cahokia is impressive, even today. Among many earthworks that remain is Monks Mound, 1,000 feet by 700 feet, and 100 feet above the plain at its highest. The labor that went into it, and the political power that saw it to completion, show that this was the work of a skilled, industrious, and organized people.

Beginning about A.D. 600, Cahokia grew for six or seven centuries—a time as long as that from the landing of Columbus to the present. Then, apparently beginning in the thirteenth century, the city declined. Even as it did, one feature of construction continued: a stockade of timbers—in fact, four consecutively constructed stockades. They enclose some of the largest platform mounds in the area, and M. L. Fowler (1975) suggests that they might have protected a "central core of the community." Might the chiefs, the rulers, who had great power, have felt a need to protect themselves? From whom?

That isn't known. Nor is it known why Cahokia declined after 1200. Perhaps the land became steadily less productive under pressure of primitive agriculture. Or perhaps, as we have suggested for Mycenae 2,400 years before, civil unrest threatened the established power. Perhaps the rains did not come, the cornfields dried up, the game became scarce. In the early eighteenth century when French traders came, they found only small scattered villages, remnants of the great mound-building people.

The indicators of climate have not been sufficiently analyzed for Cahokia. What really happened there? Our weather map for A.D. 1200 shows that when westerlies blow harder across the Great Plains, the finger of dry mountain air—the prairie peninsula climate—reaches over Cahokia.

A score of bad decades

The Mill Creek farmers, and their contemporaries throughout the plains, were not victims of just the biblical seven years of famine, or even of a human generation of bad years.

The changed climate that brought the downfall of the plains farmers probably lasted about 200 years. Our records from Europe, the plains themselves, and elsewhere show stronger westerlies between about 1200 and 1400. Then, as figure 2.1 suggests, the pattern switched back. Rains must have come again to the plains. In fact, it appears that the final decline of the Mill Creek people came just before the westerlies changed back again and more rains came to northwestern Iowa.

If they had only held out a little longer, their culture might have made it through.

Still, they didn't do badly. How would we, the civilization of twentieth-century plains corn farmers, have done?

II OUR CLIMATES SINCE A.D. 900

CHAPTER 4

One Thousand Years in Iceland

AS THE FLOW of westerlies around the North Pole expanded in the twelfth and thirteenth centuries A.D., bringing drought to the Mill Creek people, and damp winters in western Europe, what happened in other parts of the Northern Hemisphere? What happened, for instance, in the North Atlantic—where ice moving southward from the Arctic meets the Gulf Stream?

An expansion of the westerlies—in effect, an expansion of the Arctic—should bring sea ice farther south. Conditions should have changed measurably in the border region between arctic and Gulf Stream waters. This is an area as sensitive to climatic change as the Iowa ecotone between tall prairie and steppe. The North Atlantic, Greenland, Iceland, would be another place to take the measure of climatic change, to see if we understand its mechanisms and dimensions.

The object of our research on Mycenae and Mill Creek was not simply to solve ancient mysteries. We wanted to know how quickly

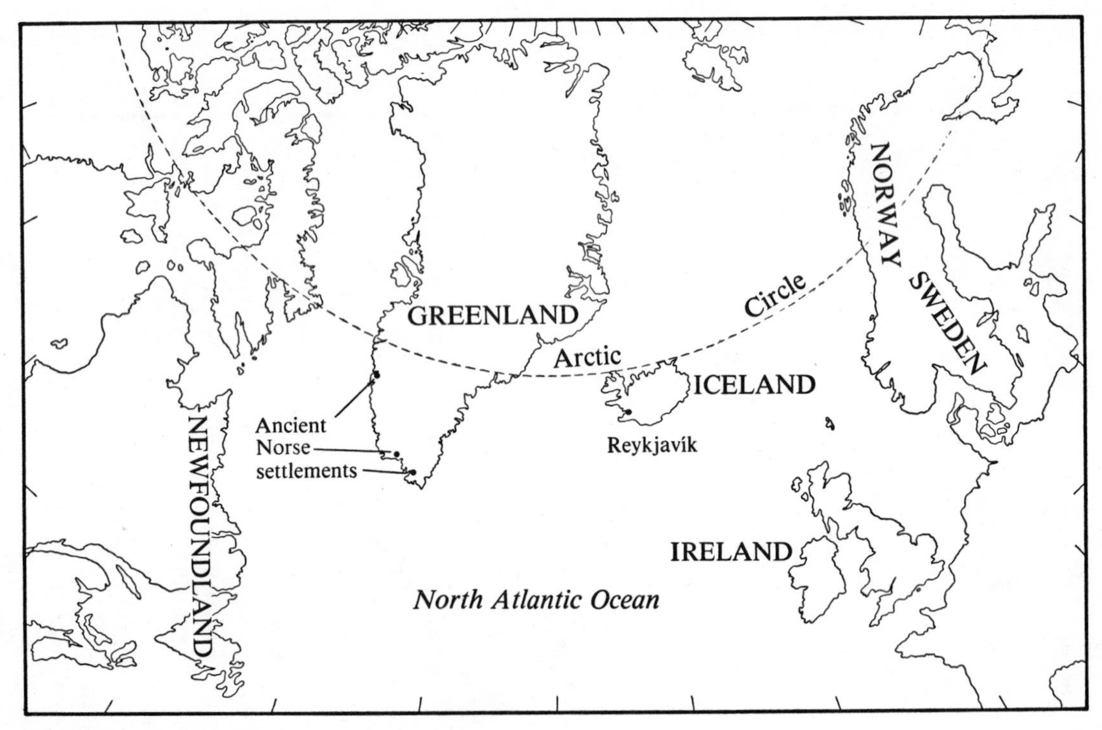

Figure 4.1. The North Atlantic.

climate could change, and by how much. And we wanted to know the causes of climatic change.

Iceland, about the size of Indiana, lies just south of the Arctic Circle. Fish and agriculture, mostly herding and wool production, are its staples. Despite its high latitude, the Gulf Stream makes possible the growing of potatoes, turnips, hay, and in certain times barley and other grains.

About 750 miles to the east lies Norway. Greenland is a like distance to the west (see figure 4.1).

Iceland is of particular interest to us because it has a recorded history of more than 1,000 years, chronicles not so much of the doings of armies and diplomats as of the lives of individual Icelanders and families.

The first sagas

"Saga" has come to mean any long narrative, but the original sagas were Icelandic prose narratives of the twelfth and thirteenth centuries. Today these are a source of information about climate.

In references to those times, the terms "Viking" and "Norseman" are often interchanged. If original use is followed, "Norse" applied to any ancient Scandinavian–anyone from Norway, Sweden, or Denmark, or people of Scandinavian stock who settled in Iceland or Greenland. "Viking" had a narrower meaning. The Vikings were the pirates, the raiders and plunderers of the whole northern European and North Atlantic area for about three centuries beginning in the eighth century A.D. Early in this period the Vikings used longboats, graceful and useful in coastal waters, but not built for the open seas. Later they adopted the Norse knorr, stubby in appearance but extremely seaworthy. Although a widespread belief exists that a knorr could sail only with the wind, these ships, in fact, did very well both ways.

The North Atlantic lands, including Iceland and Greenland, were settled by a mixture of Irish, Norse, and other Europeans trying to escape the Vikings, and by the Vikings themselves.

In 1914, Otto Pettersson published a paper on "Climatic variations in historic and prehistoric time." Pettersson drew on the sagas and on recorded indications of climate from the time of the Vikings until his day.

Pettersson reported:

> The earliest information we possess regarding the climate of Iceland is derived from the record of the monk Dicuil of Ireland in 825. He describes a visit some 30 years earlier by some Irish ecclesiastics to the Island of Thyle (Iceland). At that time, about a century before its colonisation by the Norsemen, Iceland was visited and inhabited by the Irish. The sagas call them "Papar," which indicates that they were monks or hermits. . . . Dicuil narrates the description of the island given by his fellow monks, who had been there from February to August. . . . [P. 7]

Dicuil's report, as Pettersson quotes it, was that Iceland was not surrounded by ice, even in the severe months of winter and early spring, but that "after a day's journey to the north they found a frozen sea." As Scandinavians began to occupy Iceland in the late ninth century, the climate remained favorable. Pettersson states that the sagas "nowhere mention that driftice hindered the norsemen in their [early] journeys to and from the island" (p. 7).

This ice-free time lasted hundreds of years, Pettersson says:

> Although weather conditions are often mentioned in the older annals and sagas I cannot find that the annual ice-drift to the shores of Iceland is spoken of before the 13th century.
>
> In the 13th century Iceland began to get blocked by driftice from Greenland. The blockade was much more severe then than now [1914], although even now the northcoast is frequently blocked and sometimes, though not often, the east and southeast coasts. Owing to the influence of the Irminger current, the westcoast in our time is nearly always free from ice. . . .
>
> The ice conditions in Greenland are intimately connected with those in Iceland. The advance of ice out of Nordbotn in the 12–13th centuries, of which Bardsson speaks, proved fatal to the old Norse colonies in Greenland because it cut off communication with their mother-country. [P. 8]

So the twelfth and thirteenth centuries were times of catastrophe not only for the Mill Creek people, but for cultures thousands of miles away. The Arctic did expand, bringing increased westerlies and dry western air to Mill Creek, and bringing icy death to Greenland.

Otto Pettersson's contribution to our understanding of climates, past, present, and future has generally gone unnoticed, in part because it happened to be written at the wrong time. The climate was especially favorable in the decades following 1914, in Iceland, Greenland, Sweden, and most of the hemisphere. Climatically, times were good; who cared if Eric the Red's colony froze back in the fifteenth century?

Pettersson's work, valuable as it is, was only a beginning. The ancient records held much more than he knew, and it remained for others to write a more detailed history of climate in the North Atlantic. The others included Thorvaldur Thoroddsen, a contemporary of Pettersson, and Sigurdur Thorarinsson, a later scientist. The work of these men has been compiled and added to by Páll Bergthórsson (1962, 1969) of the Icelandic Meteorological Office.

Tropical waters and ice

Iceland's strategic location in the meeting place of arctic waters and the Gulf Stream makes its 1,000-year record one of special interest—and in fact makes such a record possible.

It is easy to get a 100-year record of temperatures in Iceland; in fact, thermometer records there go back to 1846. But to go back further, Bergthórsson and others depend on reports of drift ice floating by the island. Drift ice is carried from the arctic ice pack and the waters north of Iceland by ocean currents. In cold times arctic waters carry the ice southward. In warmer times, the warm Gulf Stream dominates the Iceland area, keeping drift ice away. Iceland and Greenland are far enough north to observe this ice and sometimes to be blockaded by it; yet they are far enough south so that in historic times they have not been surrounded by solid ice. And in the warmer spells of historic times, little or no ice floats by them.

The drift ice is easy to see, both from shore and from ships. From early days, down to the sinking of the *Titanic*, and since then, it has imperiled ships and affected commerce, and Icelanders have long noted it. They have done so in a systematic way in modern times, in some detail back to the sixteenth century; the sagas and chronicles before that mention drift ice occasionally.

Drift ice, then, is a thermometer of the North Atlantic. But the readings must be interpreted. How do we calibrate records of drift-ice sightings, including some incomplete records, in terms of degrees of temperature? How cold was it when heavy sea ice threatened ships traveling to Norse homelands?

Bergthórsson and others have applied some clever techniques to this problem. While the answers they get are for average temperatures covering a decade or longer, and not for the daily readings we are accustomed to, the results are both fascinating and instructive.

A basic principle in the interpretation of old, indirect climatic records is that of calibration with the detailed weather records of modern times. We have already seen examples of this process. For instance, the pollen records from Mill Creek can tell about climate because we know, from current scientific observation and understanding, which plants need wet weather and which ones have an advantage in dry weather.

With drift ice, and with many other climatic indicators, modern records can be put alongside precise scientific records–in the case of drift ice, thermometer readings–from the same time. These precise records are a yardstick with which to measure the old records.

In Iceland, we know the number of months drift ice appeared, *and* the temperature, for every year since 1846. So it's possible to make a chart of temperatures based on drift-ice information.

Bergthórsson worked with the average temperature for each decade, and in the records of the past century he found a direct and rather simple relationship between temperature and ice sightings. For example, if ice is sighted in 20 months of one decade, and 22 months of the next decade, that second decade is about 0.1°C (0.2°F) colder than the first. The rule, however, must be applied carefully; it does not hold as well in warm times. That makes sense; at the point when no drift ice is sighted at all, any additional warming can't make a difference.

For the time before 1780, however, there aren't many records that discuss drift ice. The reconstruction of average temperatures in Iceland is, therefore, based on four types of information available for various times (Bergthórsson, 1969, pp. 94–95).

1846 to the present	thermometer readings
1781 to 1845	a chart of drift-ice sightings in each year, compiled by Thoroddsen
1591 to 1780	other weather information compiled by Thoroddsen from historical records (drift-ice records are sketchy)
900 to 1590	records of famines and severe years

Bergthórsson, then, had to establish the correlation between drift ice and thermometer readings; going further back he had to make some interpretations of recorded general comments on the weather; and (for the oldest period) he had to correlate severe years with drift ice (which in turn links them to temperature).

Reconstructing a decade

We can get an idea of how Bergthórsson went about this work from one decade, the 1750s, in the period for which, on the basis of general weather comments and a few comments about drift ice, he had to estimate drift ice for each year. He did so as follows (1962, pp. 3–4):

1751 No mention of drift ice [in the chronicles], but extensive "fast-ice" said to be at west coast. Comparison with the years 1866 and 1881, when fast-ice was rather similar, indicates drift ice duration of 5–6 months.
1752 No mention of ice. Cold from late February until Easter. Cold spring. The frost was so severe that the earth cracked in places. The same phenomenon occurred in 1918, when drift ice duration was 1½ months. Furthermore, average ice duration after a severe ice year like the preceding one is two months.
1753 No mention of ice. There must, however, have been some drift ice, since in 1758 it is remarked that no drift ice was sighted and that this had not happened for time immemorial. Estimate: one month.
1754 A very severe winter. Drift ice lying well into the summer. Estimate: six months.
1755 Drift ice disappeared September 3, which is very late. Very extensive ice in the spring. Estimate: seven months.
1756 Drift ice from early March until August 25. Estimate: six months.
1757 Prolonged drift ice until early June. Estimate: three months.
1758 Statement that no ice was sighted.
1759 Severe drift ice off the north coast and even reaching the S.W. coast. Under usual conditions this would mean a drift ice duration of three to seven months, but since the winter, spring, and summer were not severe, the estimate is three months.
1760 No mention of ice. Estimate one month.

With these estimates of drift ice, then, Bergthórsson applied his formula that converts ice duration to temperature.

For the time before 1591, Bergthórsson did not even have the skimpy information above. He did, however, have some references to bad years, years of famine. He says:

> There is hardly any event related to weather which is more worth recording than famine years. These are very much associated with coldness and ice in Iceland. The usual succession of events is the following.
>
> The springs and summers were cold and the hay-making failed frequently. Then a severe winter killed a considerable part of the sheep, horses, and cattle. This was possibly not serious for the richer farmers who were usually better supplied with food and hay, but the poorer farmers quite often had very limited supplies and therefore their losses were heavier. The only possibility was then to go out and beg. The number of wandering beggars increased tremendously in such years, but at the same time the generosity of others decreased at the same rate.
>
> This resulted in starvation of the lowest classes, people dropping dead on their way between the farms, of hunger directly or diseases associated with hunger. In addition to this it seems that the fishery failed preferably in cold periods. With increasing coldness the frequency of severe famine years rose at an increasing rate. [1962, pp. 9–10]

Bergthórsson, then, uses historical references to severe years to extend his temperature chart back. Of course, this requires calibration of famine years in more recent (and hence better known) times with drift ice, and therefore with temperature.

He does have a small amount of information about drift ice off the southwestern coast of Iceland from before 1591. Ice appears there only in very bad times; the main flows are along the eastern coast of the island. Since the population is concentrated at the southwest, as is the fishing industry, ice is mentioned even when not much else is.

Bergthórsson brought all this work together in the graph of figure 4.2. The solid part of the curve is taken from actual thermometer readings; the section before the mid-nineteenth century is estimated. For the early fifteenth century, Bergthórsson felt he did not have enough data to estimate temperatures.

Bergthórsson's picture shows that from the settlement of Iceland, about A.D. 900, until the late twelfth century, the times were warm,

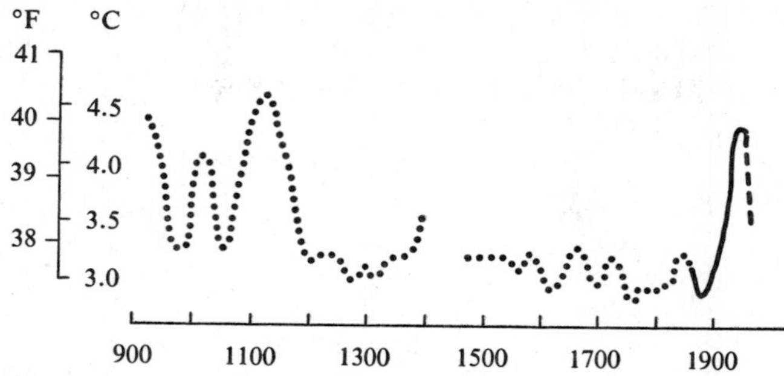

Figure 4.2. A thousand years of Icelandic temperatures. The dotted line shows estimated temperatures (with the early part of the fifteenth century omitted because of insufficient data); the solid line shows temperatures based on thermometer readings: both from Bergthórsson, 1969, p. 98. The dashed line shows the change in recent years for the Northern Hemisphere; based on Bryson 1974a.

though variable. Then Iceland was quite cold for almost 200 years. After the fourteenth century came a time of less drift ice and fewer beggars, though we are left somewhat unsure of the early fifteenth century. From the late sixteenth through the nineteenth century, the best years in Iceland were only as good as the worst years the Vikings had seen. Then, in our century, a great warming set in.

How much confidence can we place in this chart, a record of temperatures based on such incomplete information? In some cases the reading for a whole decade or more comes from just a mention or two of famine.

Precision and poetry

Such a reconstruction offends those who in this age of science and technology, of precision in measurements, expect to know rainfall to the nearest one-hundredth of an inch. We have not a single thermometer reading from Iceland before the nineteenth century. But Bergthórsson did have the drift-ice and famine records, as well as literary records like the following poem by Olafur Einarsson (1573–1659), a pastor in eastern Iceland:

> Formerly the earth produced all sorts
> of fruit, plants and roots.
> But now almost nothing grows. . . .

Then the floods, the lakes and the blue waves
brought abundant fish.
But now hardly one can be seen.
The misery increases more.
The same applies to other goods. . . .

Frost and cold torment people
The good years are rare.
If everything should be put in a verse
Only a few take care of the miserables. . . .

[Bergthórsson, 1962, p. 21]

Does such a poem convey less of reality than a temperature chart? Perhaps it gives more.

Whatever poetry does tell us, we would like to know the arithmetic of past climates. In the next two chapters, we will set a number of details against this chart. For now, we can consider it a hypothesis, a proposition to be tested. We can use it as a possible climate of the last 1,000 years, and seek other evidence to confirm or refute it.

CHAPTER 5

The Flow of Wine, Water, and Ice

HAVE THE PAST 1,000 years seen a climate essentially constant, as some people claim, or have some times been significantly different from others—not just varying from year to year, but with prolonged periods especially favorable or unfavorable to established human activity like agriculture? The question is crucial, for it bears on the possible directions of our own climates and our underfed world.

We have proposed the reconstruction of temperatures in Iceland as a record of climates for the North Atlantic since about the year 900. This chapter will describe some other kinds of evidence we have about climatic change. We will describe what indicators of past climates are available, and how they can be read; because climates around the world are linked by the westerlies we will, in chapter 6, be able to compare their records with the Icelandic temperature records of Bergthórsson.

The nature of the evidence

Most evidence of past climates is indirect. It does not consist of weather records as we normally think of them. In each case, climatologists and others must decipher the language of men and climates past as best they can. Nature does not make mistakes in the records she leaves. We sometimes do not understand them properly; that is the source of the difficulty.

Historical records from before the days of weather instruments often do not show climatic changes plain and clear. We saw the work required for Bergthórsson to reconstruct temperatures from Icelandic chronicles; Emmanuel Le Roy Ladurie, H. H. Lamb, and many others have likewise put together mosaics of the past one fragment at a time.

Trees remember

Tree rings are one important indicator of past climates. The amount a tree grows each year is related directly to the weather it experiences. The rings of a stump, or those in a thin core taken from a living tree, tell about conditions over its lifetime. In fact, tree-ring records have been extended back beyond the lifetimes of trees now living. Living trees overlap the lives of timbers in buildings, which overlap still older timbers. The overlap, the years shared by two trees in the sequence, shows the "signature" of the years those two trees shared, and makes it possible to know what years each timber came from. In Europe, work has been done on building timbers that grew in the fourteenth century.

In some parts of the western United States, tree rings take us back to the time of Christ and beyond. The old age to which pine, Douglas-fir, redwoods, and some other trees live make them ideal for looking into the past. Many redwoods live 2,000 years or more, but even these are children to the bristlecone pines, which grow high in the mountains of California, Nevada, and some other western states, without the grandeur of redwoods. They are short, gnarled, more dead than alive—often only a strip of live wood will remain on a dead trunk. Yet some of these pines are over 4,000 years old.

Just what do the rings of a tree tell? The light wood formed in the spring when the tree grows fast alternates with dark, dense wood of late summer and fall.

Wide rings show good growing years. But how that translates into climate depends on the tree and the growing conditions. In dry and sunny Arizona, rainfall is the limiting factor. There is always enough sun; when it rains the trees grow. Wide rings in Arizona trees mean relatively wet years. In the boreal forest of Canada and in other cold regions, temperature is critical. A small ring means a cold year, and if the years are cold enough for long enough, the forest boundaries shift to the south.

In climates between these two, both temperature and rainfall play a role, and sometimes the relationship is complex. Even in Arizona, where rain is the basic determinant, timing of the rain is crucial. The Utah juniper shows a small ring when precipitation has been low between October and February. And temperature, specifically temperature from March to May and in October and November of the previous year, plays a role for that tree, too.

A tree does not respond instantaneously to the weather around it. Trees store up nutrients in their roots during a good year, and a bad year's ring thus doesn't look as thin as you might expect. If a good year follows, some of its bounty makes up the deficit, leaving the ring not quite so wide as it would have been otherwise.

So the storage capacity of trees blurs the picture they give of short-term changes in climate. Those who would read past climates in tree rings have to know the way each species of tree is influenced by previous years, what information careful analysis can provide, and what is simply not available from a given source.

Tree rings also do not portray long-term changes accurately, for a different reason. Suppose 20 trees are growing on an acre of land during a time of stable climate. Then comes a drier time. During the first year or two the trees grow somewhat less than before, but they have stored up nutrients, so the drought must last a few years for rings to become much narrower. It is in this intermediate period that rings give their truest reports.

As drought continues, some trees weaken and die. Fewer new ones take their place; the acre can support perhaps 10 trees, not 20. After several decades each tree has twice the area from which its roots may draw moisture and nutrients. The tree rings are again larger—but not because rainfall has increased again.

The language of tree records is complex, and like any other is learned by observation and by practice—or what might be called trial and error. The climatologist observes tree growth under known conditions. He develops formulas that relate known rainfall and temperature with known growth, and that therefore can be applied to years before actual weather measurements were taken—the same calibration procedure used for North Atlantic drift ice.

For example, a formula has been devised for growth of Douglas-fir in Arizona. It takes into account, for a given year of growth, spring precipitation, winter precipitation, spring temperature, winter temperature, previous year's growth, and growth three years previously. Each of these factors is given a different weight; the previous year's growth is the single most important one. Growth records for a long period enable us to compute the weather conditions that must have created the sequence of rings.

Tree-ring records aren't as useful for looking at the general climates of the hemisphere over the last 1,000 years as some other records are. The most extensive tree-ring work is limited to the western and southwestern United States, where the rough topography makes for complex patterns, especially of rainfall. These intricate patterns are not well enough known to be fitted very precisely with changes in the westerlies.

In Europe, analysis of tree rings started somewhat later than it did in the United States. Although trees with the longevity of bristlecone pines and redwoods are not found there, buildings are older than in the United States and wood used in their construction gives information.

In the wine harvest, truth

Records that go back about as far, and exist in more detail for more places in Europe, are based on phenology, the study of how living things respond to the changing seasons.

Tulip blooms and the lark's song are called forth by the warmth of late winter and spring. The blooming and fruiting of plants is closely related to temperatures during the growing season; if you know what day the tulips bloomed each year, or when the wheat crop was ripe, you know pretty well what the temperature was—not for every day or week, but for the season as a whole.

But this kind of climate indicator is not self-recording, as are tree rings and pollen layers. Like drift ice for Iceland, phenology depends on written records. We do not have numerous or consistent records of gardens or wheat fields over the last centuries. But information abounds for an activity that many people find more interesting–the harvest of grapes for wine.

Throughout the vineyard districts of Europe, town and regional records tell wine harvest dates. In many cases an official commission toured the vineyards and declared the date. Beginning in the late nineteenth century, scholars compiled thousands of references to wine harvest dates. Emmanuel Le Roy Ladurie's book, *Times of Feast, Times of Famine* (1971) contains a wealth of information about trees, grapes, glaciers, and other indicators of climate since the year 1000.

Some measure of human judgment enters into the setting of the harvest dates; they are not pure reflections of climate. Leave the grapes too long and risk losing some; harvest too early and the wine is inferior. The wine market of the time has an influence, as does the opinion of local officials. But just as 100 trees tell a convincing story of climate though one tree might be suspect, these influences average out when many wine records are examined together. The result is a remarkable correlation of wine harvest dates with growing-season temperatures, as shown by figure 5.1.

Note that small temperature changes have a strong effect. If the average temperature over the summer goes up by 1°C (1.8°F), the wine harvest comes about ten days earlier.

Of course, wine harvest dates have limitations. They do not reflect climatic trends that take place over a century or more, because vintners change their varieties, and strains of each variety, according to what is most productive for them. If the growing season becomes slightly cooler, harvest dates slip back somewhat. The harvest proceeds, though the wine may not be outstanding. But in such a cooler time, fall frosts come earlier, too. If the cooling persists for decades, the wine-growers gradually change to grapes more suitable for the new climate. Over two or three human generations a new "normal" is accepted–grapes that ripen during whatever growing season is then available. And if a region's "great" wine is really not up to the standards of great-grandfather's day, few comparisons are made anyway.

Still, wine harvest dates do show shorter-term climatic changes, and they show the onset of long-term changes. In the next chapter

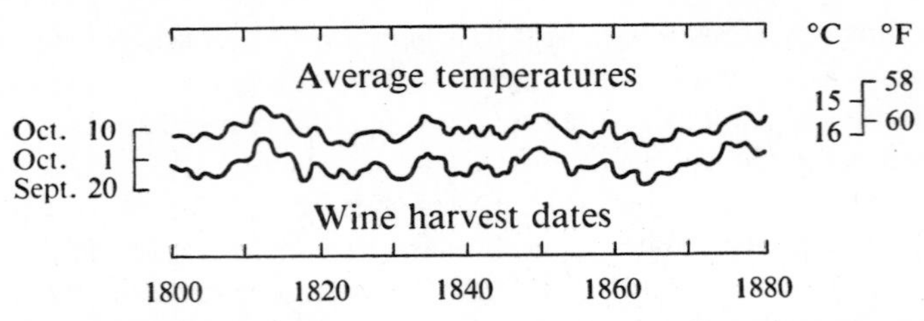

Figure 5.1. Comparison of French wine harvest dates, and growing season average temperatures from thermometer readings. Adapted from Ladurie, 1971, p. 51.

we will see what they tell us about the past 1,000 years. As with any record of climate, they must be used with caution: one should not read into them more than they contain.

What did Ben Franklin know?

Suppose the area under arctic air expands—the circle of the westerlies grows larger, and Iceland feels the chill. Drift ice moves farther south, and the whole North Atlantic cools off. The Gulf Stream, which in warmer times brings water from the Caribbean along the east coast of the continent and up past Iceland, would be affected too.

The Gulf Stream is part of the general clockwise rotation of currents in the North Atlantic. It flows from Caribbean to Atlantic through straits between the Florida Keys and Cuba, roughly parallel to the eastern coast of the United States up to Cape Hatteras, then bends eastward. Eventually currents reach Europe, and move down past Spain and North Africa, to return westward and into the Caribbean again.

The Gulf Stream was discovered about 1400, and since then sailors and map-makers have taken note of its importance to Atlantic traffic. But old records differ as to the stream's course between Cape Hatteras and Europe. Does it turn directly eastward and flow toward Portugal? Does it go north-northeast, to Iceland and beyond?

Despite the odd appearance of maps from the seventeenth century and before—the shapes and sizes not quite right—seamen and navigators from the Vikings on were very well able to set accurate courses between Europe and North America. From their maps and the descriptions they gave, it is clear they knew the ocean currents in detail.

In the late 1760s Ben Franklin turned his formidable attention to the problem of the Gulf Stream. More than his own scientific curiosity was at stake: he was deputy postmaster general of the colonies, and English mail packets were taking two weeks longer to make the westward crossing than were Rhode Island merchant ships. Sea captains from the colonies told Franklin that this was because they avoided the Gulf Stream on westward crossings while the English captains did not. Information from these captains, and from his own temperature readings on Atlantic crossings, enabled Franklin to draw Gulf Stream maps (Stolle, 1975).

The maps of Franklin, English voyager Martin Frobisher in the sixteenth century, and others do not agree with today's maps. This has been chalked up to ignorance on their part. In the next chapter we will see whether their descriptions of the Gulf Stream's course tally with the climate reconstruction of Bergthórsson and with other data we have from the past 1,000 years.

Glaciers waxing and waning

North America and Europe have not been covered by the continental glaciers of an ice age for 10,000 years. But in Scandinavia, the Alps, and elsewhere are reminders of colder times. Mountain glaciers extend their fingers of ice down valleys and into the habitat of man, to a greater or lesser extent according to temperature and snowfall.

In populated countries, the movement of glaciers has attracted enough attention to be recorded during the past several hundred years. Certainly, in the extreme case when a glacier advances right through a mountain village, notations are made in the tax records, if nowhere else. Church records too show when priests and bishops were called to bless the fields of Alpine villages and pray that the ice stop grinding forward. Hundreds of such records give us a picture of glacial movements over the centuries.

We saw that neither tree rings nor wine harvest dates give an accurate picture of long-term climate trends; adaptation in both cases changes the rules of the game. But glaciers march to a slower drumbeat. They are massive and can change by only a small percentage each year, even if weather conditions change abruptly. It's the average climate of decades or centuries that determines how much they grow or shrink.

Glaciers, like most other indicators of climate, respond to several kinds of changes. They don't indicate only temperature, or only snowfall. Winter is their growing season. They feed on snow, and winter precipitation is one of the main determinants of their progress. But winter temperature is not: as long as snow does not melt, coldness doesn't add or subtract. In fact, very cold winters over a large area might mean less air moisture and hence less snow.

In summer, especially at a glacier's borders, some snow and ice melt. Glaciologists call this process ablation, and it is just as important in determining the glacier's course as is snowfall. For the glacier to maintain its position, snowfall must equal snowmelt, so that mass remains constant. If snowfall is an inch greater than melt, the glacier must grow. Glaciers are long-term balance sheets. A few inches of nonmelting snow a year for a few decades will advance a mountain glacier far down into the valleys. A few feet of snow a year for centuries makes an ice age.

Despite the rather involved relationship of temperature and precipitation in the formation of glaciers, their growth does correlate well with long-term average temperature. And they are quite sensitive to it. A small temperature change will tip the balance toward more ice or less (Ladurie, 1971, p. 230).

In the ice age that ended 10,000 years ago, globally averaged temperatures were probably only about 5°C (9°F) colder than now—and the winters were not more severe than ours and may have been a little warmer (Bryson, 1974a; Moran, 1976). In a given day or week, such a change would hardly be noticeable. Canada and the northern United States were by no means engulfed in a continual, howling blizzard. Yet across Canada stood a vast mountain range of ice, up to two miles high.

In our own times, the advance and retreat of Alpine glaciers have come with temperature changes of a degree or two Fahrenheit. This is enough to move a mountain of ice, and quite enough to move the affairs of humans.

CHAPTER 6

The Past 1,000 Years: Europe, the North Atlantic, the United States

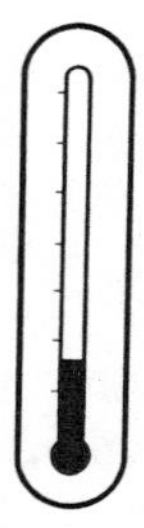

AS WE LOOK AT FACTS about the past 1,000 years, the component of climate to which we will pay most attention is temperature. Yearly average temperatures are a better indication of climatic changes than precipitation or other aspects of climate. While temperatures do not all change at the same time or in the same amount, they do show general climate trends. They must: changes in the westerlies bring changes in the boundaries between arctic air and warmer air. Precipitation, on the other hand, is more spotty and variable, more influenced by geography. Precipitation records are not so useful for showing the broad directions of climate.

It's well to keep in mind, however, that averages are far from the whole story. It is the sequence of weather events that composes "climate," not merely the average weather. Drought in one state and flood in the next might average out to "normal" for the region, as might a dry year followed by a wet one. But when the extremes of weather become greater, an average figure hides the true nature of things. The destruction such sequences bring should not be overlooked by considering only average readings.

Still, generalizations are a place to start. If we know there is a general expansion of the westerlies, we can move on to examine the resulting climate changes in England, the Great Plains, or elsewhere.

Our point of reference: Iceland

Keeping in mind the westerlies as the link between regional climates, we will see how Bergthórsson's reconstruction of Icelandic temperatures (figure 4.2) holds up in the light of what else we know about climate since A.D. 900, paying particular attention to the indicators described in chapter 5.

Most of our past climate indicators, including Bergthórsson's graph, paint their pictures with a rather broad brush. We often can't see the details of individual years or even individual decades. It would make no sense to seek close correlations for short periods of time. The complexity of climates and their variation from locality to locality are other reasons to look at broad trends.

Looking at Bergthórsson's chart, we divide the 1,000 years into four general periods, as follows:

> *First,* 900 to 1130, a warm though variable time in Iceland;
>
> *Second,* the cooling that began in the mid-twelfth century and lasted until about 1380;
>
> *Third,* the even colder time from 1550 to 1850;
>
> *Fourth*, the warming from then until the middle of our own century.

Dates, of course, are approximate, but we use specific years as beginning and end points for convenience. The century or so around 1400-1500 appears to have been an unsettled time; it is one for which we lack information for Iceland and for other places, and we have, therefore, omitted it.

900-1130: Age of the Vikings

The Scandinavian conquerors—pirates, navigators, and settlers—symbolize this time, and perhaps 100 years or so on either end of it. But their coming-out was not alone. The North Atlantic and North

Sea regions came alive as in a springtime. The Celtic church in Ireland sent missions to Iceland and to Africa. In the late eleventh century, dozens of vineyards produced wines in England, where today none are successful. Some belonged to the king, who himself had crossed from Normandy two decades before. (See Lamb, 1966, p. 7.)

But the Norse became the world travelers. They settled Iceland and Greenland; Eric the Red's colony in the latter began in the tenth century. They made voyages to North America. They traded with Italians and Arabs, and they traveled the rivers of Russia (Herrmann, 1954, p. 318). In their homeland, too, the limits of settlement expanded. Forests were cleared and farming moved farther up the valleys and hillsides of Norway (Lamb, 1966, p. 174).

Various explanations are offered for all this activity. Perhaps it was technologically based, coming from more use of iron and other metals. Perhaps it was a search for new land and resources to satisfy a growing population. Some of the pioneers were trying to escape Viking raids or harsh rulers. One force behind the movement to Iceland in the ninth century was the ruthlessness of Harald Fairhair, a Norwegian king.

Whatever moved the Scandinavians to break out from their homelands, they certainly had nice weather for it. Perhaps that was reason enough. The climate, Bergthórsson's chart suggests, was about the most favorable of the past 1,000 years. A variety of forces, however, combine to shape the lives of people and their nations. The settlement of Greenland had to take place in a mild North Atlantic, but society's rejection of a couple of brawlers had a hand too (Herrmann, 1954, pp. 246–254).

In 960 Thorvald Asvaldsson of Jaederen in Norway, a violent, quarrelsome fellow, killed a man. He had to leave the country; he took his family to Iceland. They went to the northern part of that island because the more pleasant areas were already settled. He had a ten-year-old son, Eric, later to be called Eric Röde, or Eric the Red. Eric, when he grew up, married into a good family and got himself a better farm. But Eric too had a violent streak. He killed two men, and in 982 was banished from Iceland for three years. He made good use of the time.

Eric sailed west to learn about a land that Icelanders had discovered years before but knew little of–Greenland. He scouted out the rough coastline and found the most hospitable area, a deep fjord on the southwestern coast. Warmer Atlantic currents met the island there, and conditions were not much different from those in Iceland.

Eric was not only an explorer, he was a real-estate developer, and he "believed more people would go thither if the country had a beautiful name," according to one of the Icelandic chronicles (Herrmann, 1954, p. 249). But Greenland was never very green. It is too far west to catch much of the Gulf Stream, and so, unlike Iceland, it had, even then, few trees and was poor farming country. Nevertheless, Eric drew settlers, who grew vegetables and hay, and lived mainly on livestock farming. At one time, perhaps 3,000 people in Eric's colony had settled 280 farms.

Eric's own homestead, uncovered by expeditions within the last 100 years, had four cowsheds with 40 stalls. And in our century, although it is warmer than many of the past 1,000 years, that place is barren tundra, and many fjords are blocked by the advance of inland glaciers.

We know of Iceland and Greenland from Icelandic sagas and histories, especially the *Landnámabók,* or *Book of Settlements.* This detailed record of the discovery and settlement of Iceland was begun in the twelfth century but includes material from several centuries before. In the old Icelandic records is also described the sailing route from Iceland and Norway to Greenland. For the earliest times, until the fourteenth century, the directions included the landmark Blaserk, "black mountain or peak," somewhere near Cape Farewell at the southern tip of Greenland.

The warm Atlantic also encouraged variants of agriculture in Europe which have not been possible since. The vineyards in England were one remarkable example. William the Conqueror's census of all landowners and their possessions in his newly conquered kingdom, the Domesday Book (1085), records 38 vineyards in England, besides those of the king. These ran as large as ten acres; five of them operated for more than 100 years, and the wines were said to be comparable with those of France (Lamb, 1966, p. 7).

Whether the wines were good or not, the vineyards did produce. H. H. Lamb has mapped them, and compared the present climates in those places to climates at the northern borders of wine production on the continent. In all cases, the May or July temperatures–or both–in England fall short of what is needed. Lamb adds, "It is true that individual enthusiasm has succeeded in operating isolated vineyards in specially favourable sites in the south of England in most centuries since the Middle Ages, but these have never continued long after the retirement or death of the enthusiast" (Lamb, 1966, pp. 190–191).

In this time of vineyards in England and Viking dominance of the North Atlantic, Eric the Red's people found a good, if demanding, life in Greenland. Twenty-five hundred miles to the southwest, on the plains of what are now Iowa and Nebraska and Kansas, times were even better.

1130–1380: Death at Mill Creek and in Greenland

The expansion of the westerlies in the twelfth century brought drastic change: it ruined the English vineyards, covered the remnants of once prosperous Indian villages with dust, ended the heyday of the Vikings, and brought death to the colonies in Greenland.

About this time the boreal forest boundaries in Canada began to retreat from their far-north position of the previous centuries. In Europe the Alpine glaciers advanced, moving down valleys and across land people had claimed. We can read this advance in the layers of glacial gravel that alternate with the remains of plants, in the areas the glaciers cover and uncover over the centuries. This glacial advance, the one we connect with drought at Mill Creek, came about 1215–1350 (Ladurie, 1971, pp. 248–250). Given the slowness of glacial change, these dates fit well with our ideas about Mill Creek.

From tree rings, we have more exact dates for climatic changes that destroyed Indian communities in the American Southwest. At Mesa Verde in Colorado, and other settlements throughout that part of the country, the cliff dwellers died or moved on during extremely dry years from 1271 to 1285 (Ladurie, 1971, pp. 26–28, 35).

There is some evidence, from tree rings and soil erosion, that after a dry period throughout the fourteenth century, the Southwest had heavy rains for many decades. The Indians once grew corn on the mesas above their cliff homes. They could not do so today: the soil has washed down into the valleys. Heavy rains are the obvious explanation for this, but perhaps not the right one. After a severe drought had killed off most of the vegetation–trees and grasses as well as the crops–even moderate rains could have eroded the land.

The information from the American Southwest further confirms that the thirteenth and fourteenth centuries were times of climatic change.

For a climatic record of the North Atlantic, we do not have tree rings to rely on but we do have written accounts of the times. It appears that the sailing route from Iceland to Greenland had to be changed as drift ice floated down to block Greenland's eastern coast. Ships had to head farther south, then swing back to reach the southwestern coast settlements. Ivar Bardsson, a Norwegian priest who lived in Greenland from 1341 to 1364, wrote: "From Snefelsness in Iceland, to Greenland, the shortest way: two days and three nights. Sailing due west. In . . . the sea there are reefs called *Gunbiernershier.* That was the old route, but now the ice is come from the north, so close to the reefs that none can sail by the old route without risking his life" (Ladurie, 1971, p. 253).

About this time, references in the ancient chronicles to the Greenland landmark used by sailors became confusing–was it a black mountain or a white mountain? Fridtjof Nansen, the modern Norwegian explorer and geographer, did not believe climate had changed within historical times. He struggled with the references to that landmark:

> It is more difficult to explain the two names Blaserk and Hvitserk [literally, "black shirt" and "white shirt"] which were the most frequently mentioned especially later on. They have often been confused with one another, and while Blaserk is mentioned in the oldest records, Hvitserk gradually supplants it in later writings. Later authors often mention the names in an opposite sense, Blaserk representing a dark glacier or mountain peak, Hvitserk representing a white one.
>
> It is a curious fact, that, while Blaserk is mentioned only in the older writings, such as the Landnama and Eric Raude's saga, this name almost disappears from later writings and is supplanted by that of Hvitserk, which name is first mentioned in manuscripts from the 14th century and later. . . . In no Icelandic manuscript I have found both names used simultaneously, it is always one or the other, nor are they ever mentioned as representing different localities on the Greenland coast. It then appears too rash to conclude, as hitherto, that the names indicate two mountains. [Quoted in Pettersson, 1914, p. 13]

Nansen had a convincing case that Blaserk and Hvitserk were one and the same, but he couldn't solve the confusion. He faced the same perplexity as those who argued that Mycenae was certainly overtaken by invaders, but then proceeded to show all the routes the invaders could not have taken.

Otto Pettersson (1914, p. 12) drew the obvious conclusion: "... Blaserk ... was so called in the time of Eric Röde because it then was free from ice and snow, which a few centuries later covered it and changed the name to Hvitserk."

After a time the landmark, whether a black mountain or a white one, was irrelevant, for no ships sailed to Greenland. In the worsening climate, voyages were more difficult and less profitable, and were not undertaken by Icelanders and Norwegians, who had their own troubles closer to home. The graves and ruins in Greenland show the people there made an attempt at civilized living to the end, with their rough clothes sewn in the style of the day as nearly as they could determine from a few outside contacts. But the cold and the lack of proper nourishment took a terrible toll. The early Norse settlers stood 5′7″ or taller (Bergthórsson, 1962, p. 11), but by about 1400, Lamb says (1966, p. 8), the average Greenlander was probably less than five feet tall.

After World War I, Denmark sent a commission to Greenland which found remains of the early settlements. In their last years the Greenlanders were severely crippled, dwarflike, twisted, and diseased (Herrmann, 1954, p. 244). Some historians believe that Eskimos, perhaps driven from the north by the colder times, attacked and killed some of the settlers. In any case, Eric the Red's colony survived for 500 years, ending in a slow and painful death. Those people lie buried in ground now permanently frozen, and frozen too in that ground are the roots of trees able to live in Eric's Greenland (Lamb, 1966, p. 188).

1550–1850: Only a little ice age

In the times we have just discussed, the thirteenth and fourteenth centuries, the evidence available suggests that the westerlies had made a slight shift to the south. But for the most part, they swept in a west-to-east circle, without broad north and south looping.

The fifteenth century, however, saw some decades when the westerlies expanded even more, and others when they shrank considerably. This conclusion stems partly from the variation within records from that time, partly from the seeming inconsistency between sets of records and the disagreement among scholars about what happened. When people cannot agree about what a climate, past or present, is like, the chances are that the evidence supports either view. Farmers on the Kansas plains in the mid-nineteenth century argued whether their climate was normally benign or normally hostile for agriculture. In fact, it brought great years and terrible ones; its obvious characteristic was variability. The 1400s were such years throughout the hemisphere.

Then, in the sixteenth century, the expanded and looping patterns of the westerlies came to dominate. As always, the weather varied a good deal. But the cold years were colder and more frequent; the warm years were often not as warm. Food production was often reasonably good, occasionally very good. But more often than before or since, it was bad.

It's not clear just when this cooling set in. Of course, both timing and severity varied from place to place. Clearly something had happened by the 1590s. The ponderous glaciers had advanced toward the villages in the Alps, a consequence of snowy winters and cold summers. Considering the time it takes for glaciers to reflect even the most abrupt change in their environment, 1550 is probably an accurate date for the beginning of climatic change. Even by the mid-sixteenth century, some glaciers in Europe were already more advanced than in our times. One of these was the glacier at the head of the Rhône River.

The Rhône flows from the Swiss Alps westward, then south through eastern France to the Mediterranean, west of Marseilles. The glacier from which it springs, the *Rhonegletscher,* was described in 1546 by the German cartographer and geographer Sebastian Münster. Ladurie, in *Times of Feast, Times of Famine,* relates Münster's account.

Münster rode on horseback until he came to

an immense mass of ice. As far as I could judge it was about two or three pike lengths thick, and as wide as the range of a strong bow. Its length stretched indefinitely upward, so that you could not see its end. To anyone looking at it, it was a terrifying spectacle, its horror enhanced by one or two blocks the size of a house which had detached themselves from the main mass. A stream flowed out of it, a mixture of water and ice, which I could never have crossed on horseback without a bridge. [Ladurie, 1971, p. 130]

Ladurie converts the pike lengths and range of a strong bow to dimensions of 40–50 feet high and over 600 feet wide at the front wall. He himself retraced Münster's journey in 1962. He saw that "the present glacial tongue is very thin . . . It has withdrawn high up into the escarpments and the steep and slippery rocks of the mountainside." Ladurie found that today it is impossible to ride a horse to the front of the glacier, the very beginning of the Rhône, because that beginning is now well up into steep granite canyons (1971, pp. 130–131).

We don't know what the *Rhonegletscher* had been like in the time before 1546; we don't know whether it had been growing or shrinking in the decades before that. But we do know that, throughout the Alps, the glaciers grew after that. They brought terror and death to villagers who had believed that glaciers, and climate, do not change.

The tithe was reduced

The ice came down on the village of Chamonix, in the French Alps south of Lake Geneva, about 1600. This was significant not only to the people living there, but also to the tax collectors, and their records describe what happened:

At the time of the tallage reform [this tax was reformed in 1600], the glaciers of the Arve and other rivers ruined and spoiled one hundred and ninety-five *journaux** of land in diverse parts of the parish [Chamonix] and in particular ninety *journaux* and twelve houses were ruined in the village of Chastelard of which only a twelfth part was left. The village of Les Bois was left uninhabited because of the glaciers. . . . In the village of La Roziere and Argentier seven houses were covered by the said glaciers, whose ravages continue and progress from one day to the next. . . . Because of all this ruin the tithe was greatly reduced. [Ladurie, 1971, p. 133]

*A *journal* was an old French measure of land–literally, the amount a man could work in a day (*jour*). It varied in extent from region to region.

The bureaucrats of 1600 saw their tax base eroded; to the villagers, erosion was deeper.

Throughout the Alps, glaciers thrust forward in the late sixteenth and early seventeenth centuries. This began a "glacial high tide," as Ladurie describes it, of nearly 300 years. The glaciers moved forward and back within this time, but were always more advanced than before or afterwards (Ladurie, 1971, p. 220). The "waves and swells" within this high tide brought several more episodes of destruction in the Alps. The valley of Chamonix saw more trouble in the 1640s, when officials reported:

the glacier Des Bois advances by over a musket shot every day, even in the month of August, toward the said land. . . . We have also heard it said that there are evil spells at work among the said glaciers, and that the people, last Rogation-tide, went into procession to implore God's help to preserve and guarantee them against the said peril. . . . the people sow only oats and a little barley, which throughout most of the seasons of the year is under snow, so that they do not get one full harvest in three years, and then the grain rots soon after, and only a few poor people eat it, and for sowing they generally have to go and buy other seed, and we observed that the people there are so badly fed they are dark and wretched and seem only half alive. [Ladurie, 1971, p. 170]

Perhaps some of those wretched souls were victims of St. Anthony's fire. In such conditions, even had people known the disease came from grain picked green and rotting, they might still have eaten it, and prayed for deliverance.

Within the glacial tide that lasted well into the nineteenth century, other major advances of ice came in the late 1810s and in the 1850s. Percy Bysshe Shelly spent the summer of 1816 in the Chamonix valley, and in his poem "Mont Blanc" wrote:

The glaciers creep
Like snakes that watch their prey, from their far fountains,
Slow rolling on; there, many a precipice,
Frost and the Sun in scorn of mortal power
Have piled: dome, pyramid, and pinnacle,
A city of death, distinct with many a tower
And wall impregnable of beaming ice.
. . . The race
Of man flies far in dread; his work and dwelling

Vanish, like smoke before the tempest's stream,
And their place is not known. . . .

For the later times a variety of records exist, but in some places records were no longer even kept. People had left the worst areas, including some parts of the Chamonix valley. Little or nothing remained there to destroy; the high tide had become a "normal" tide.

In our own century, the normal is something quite different. Ladurie observed in the early 1960s that the Chamonix villages once hemmed in by ice now have at least a kilometer of woods, moraines, gorges, and rocks between themselves and the glaciers (Ladurie, 1971, p. 172).

Throughout the Alpine regions of Europe, legends, customs, and stories can be traced back to the beginning of the little ice age. Elmar Reiter of Colorado State University told us a story he is personally familiar with, of gold mines in the Hohe Tauern mountains of the Austrian Alps, along the southern border of the present province of Salzburg:

The mines were operated by the Archbishop of Salzburg who was one of the wealthiest dukes in the empire. Tools found in the mines after the glaciers receded during the recent warm period would suggest a sudden and unexpected abandonment [around the onset of the late-sixteenth-century climatic change]. The mouths of the mines had been buried under ice until recently, and the tools were well preserved.

The mining tools left in the mines were supposed to be used during the following summer. It would appear that a succession of two to three bad summers, during which the mines could not be operated, would cause such economic hardship among the miners that they would have to look for other jobs. Even though it would have taken a good number of years before the mines were actually covered by ice, the economic disruption caused by several bad years in succession might have put an abrupt end to the mining operations.

Among glaciologists and historians of climate, 1550 to 1850 has come to be called the "little ice age." The name fits for the Alps. But what about other regions?

While the people of Chamonix struggled to raise oats and barley, which did not ripen if they grew at all, Icelanders faced the same problem. Bergthórsson says grain was grown in most of Iceland from

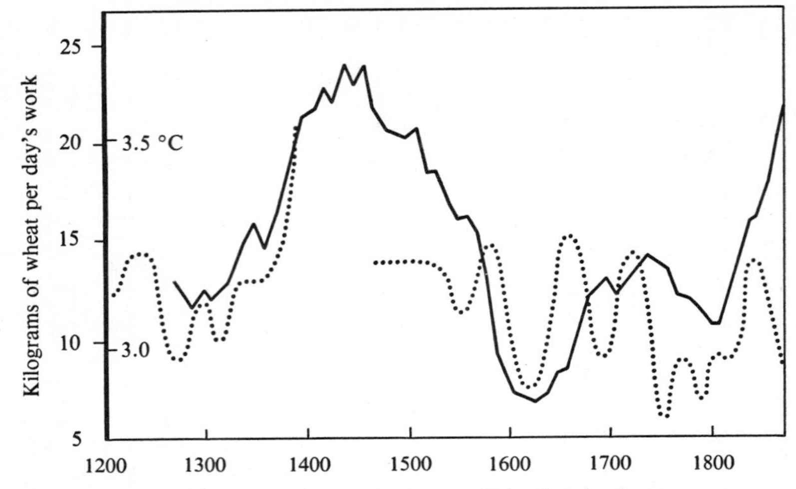

Figure 6.1. Comparison of price of wheat and North Atlantic temperatures. Solid line shows kilograms of wheat that could be purchased with a day's wage for an English carpenter. Dotted line shows North Atlantic temperatures, with the early part of the fifteenth century omitted because of insufficient data. Adapted from Martell, 1976; based on Bergthórsson, 1969, p. 98, and Libby, forthcoming.

the time of settlement in about 870. But in the cooling of the thirteenth and fourteenth centuries, only barley (a short-season grain) was grown, and only in southern Iceland. Plantings increased a bit in the next two centuries, but in the sixteenth century grain-growing was abandoned. After 1920, oats and barley again grew well and cultivation increased rapidly (Bergthórsson, 1962, p. 13; 1969, p. 100).

In regions marginally cool for grain, whether Iceland or the Alps, a small increase in cooling will have a major effect. As Bergthórsson says (1969, p. 99) about his temperature graph, "the so-called little ice age . . . is a little cooler than any other during the last 1,000 years. But it is not nearly as marked as one might have expected. It must though be remembered that a certain cooling may have more serious economic consequences in a cold period than in a relatively mild climate."

Since grain is so important to human survival, we would like to have detailed records of grain harvests over the centuries. We don't, although figure 6.1 shows that wheat prices in Europe can be correlated with North Atlantic temperatures. But for detailed, year-by-year records from several parts of Europe, wine harvest dates are best. As we saw earlier, they do match up well with the warmth of the growing season.

While wine harvest dates show only the summer weather, and glacial advance results from that and winter snows as well, these two types of records agree well throughout the little ice age. Ladurie reports notable series of late wine harvest dates in 1591–1602 and 1639–43. These correspond directly with the glacial peaks about 1600 and again in 1643–44. The glacial peaks come toward the end of the periods of cool summers, because glaciers build throughout a cool spell, and are largest at its end.

Earlier we said wine harvest dates do not show long-term climatic trends—those lasting a century or more—as glaciers do. Vintners select the grapes that grow best, and, as climate changes, the selection changes too, to get the most out of the available growing season.

A little warmth, much cold

The little ice age, like any other time, had some warm spells as well as cold ones. For instance, many summers between 1650 and 1686 were hot. In 1666 wine harvests in northern France were an average of nine days earlier than for the seventeenth and eighteenth centuries as a whole. Across the Channel the weather that summer was also dry and hot. At the beginning of September the great fire of London broke out; it raged through the city for five days, with, as Samuel Pepys described it in his diary, "everything, after so long a drought, proving combustible, even the very stones of churches"; and it destroyed some 14,000 buildings and left 200,000 people homeless.

But the cold years were even more destructive. The eighteenth century brought some terrible winters; these generally do not show in wine harvest dates. The winter of 1709, says Ladurie, killed as many people in France and elsewhere as might have died in a major war. A priest in Angers, in west-central France, wrote:

> The cold began on January 6, 1709, and lasted in all its rigor until the twenty-fourth. The crops that had been sown were all completely destroyed. . . . Most of the hens had died of cold, as had the

beasts in the stables. When any poultry did survive the cold, their combs were seen to freeze and fall off. Many birds, ducks, partridges, woodcock, and blackbirds died and were found on the roads and on the thick ice and frequent snow. Oaks, ashes, and other valley trees split with cold. Two thirds of the walnut trees died. . . . Two thirds of the vines died. . . . No grape harvest was gathered at all in Anjou. . . . I myself did not get enough wine from my vineyard to fill a nutshell. . . . Fortunately God in His infinite mercy gave Normandy cider in abundance in 1709, and it was transported here and sold. [Ladurie, 1971, p. 91]

In March, the poor rioted in several cities to keep the merchants from selling off the little wheat left. The abundance of cider must have been a comfort to many—but not to the thousands who starved or froze to death.

In 1739–40 there was another terrible winter, especially in Belgium. After it came "no spring, bad weather up till the middle of May, a short, cold, dull summer, a late wheat harvest spoiled by rain, wine harvests and fruit destroyed by the early frosts." Again the poor people rebelled; the governor of Liège told the well-off to "fire into the middle of them. That's the only way to disperse this riffraff, who want nothing but bread and loot" (Ladurie, 1971, p. 92). A theme, perhaps, from Monks Mound and Mycenae—and the future?

These winters are in keeping with the generally cooler conditions of the little ice age. Years like 1709 and 1740 were worse than average even for those times, but there were many other bad ones. The year-to-year variations in climate were variations around a cold average. Colder than average in those centuries was very cold indeed.

In England, we cannot peer at the little ice age climate through the keyhole of wine harvest dates. The success of vineyards there for decades and longer about A.D. 1000 seems remarkable even today; during the little ice age they were out of the question. But we have some English records that are even better. The development of science and scientific instruments in the seventeenth century produced quite reliable temperature readings in central England from 1680 onward. While methods of taking these readings (exposure of instruments and the like) were at first somewhat out of line with modern standards, British climatologist Gordon Manley has standardized them so we have an accurate picture of almost 300 years in a limited area (Manley, 1953, 1974; Lamb, 1966, p. 175).

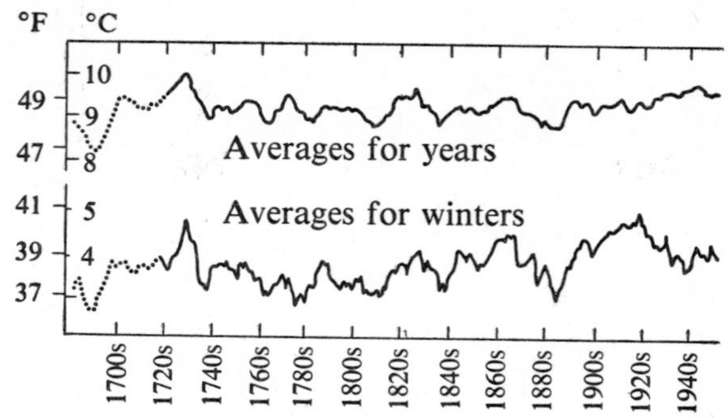

Figure 6.2. Temperatures in central England, 1680–1960. The curves are constructed of running ten-year averages, a technique that eliminates extremes of individual years. Adapted from Manley, 1953, and as quoted in Lamb, 1966, p. 175.

In figure 6.2 are two sets of readings Manley compiled, those for winters and those for yearly averages. The extremes of individual years do not appear, because this graph is made up of ten-year averages—a technique that gives a better picture of general trends, and still shows peaks and valleys that lasted only a few years.

Compared to the twentieth century, winters from the beginning of the records until the mid-nineteenth century were generally cold. In the 1910s, winters were 1.7°C (3°F) or so warmer than for many years in the eighteenth and nineteenth centuries. Winter temperatures dropped off a little toward the middle of our century, but other seasons became warmer. Summers in the 1930s and 1940s were 0.6°C (1°F) or more above the readings from the beginning of the century. The graph of yearly averages shows a general rise from the mid-nineteenth century onward to 1950.

What do these changes mean? Again we find, within the past 1,000 years, decades or centuries when temperatures moved upward or downward a degree or so. It seems like very little, judged by the much greater day-to-day variations within any time. And certainly the variations are small enough to give some people the impression of no change: England has seen no advance of either glaciers or palm trees.

Yet we should keep in mind the rather small temperature change that produces an ice age: 6°C (10°F) or less. Not startling either, if compared to typical day-to-day changes. H. H. Lamb points out that in the warmer times of the last 1,000 years, southern England had

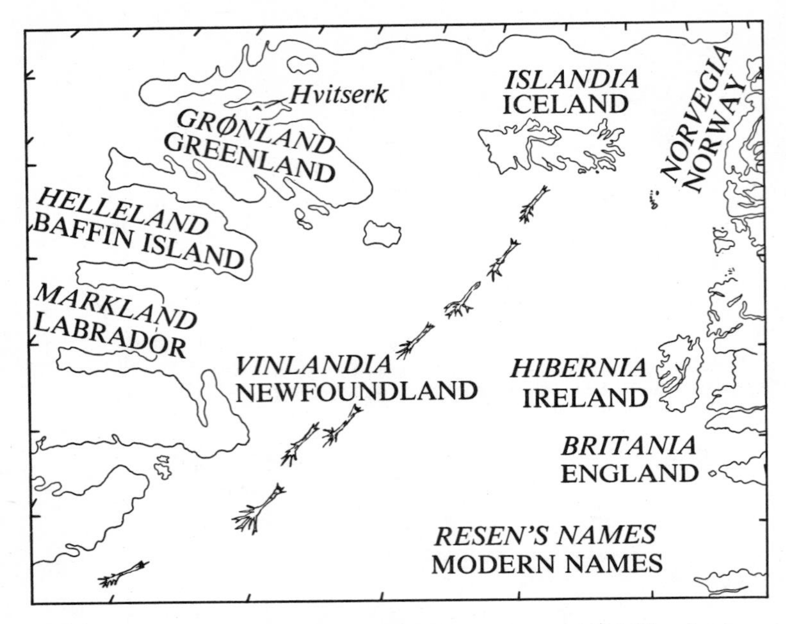

Figure 6.3. Simplified version of Bishop Resen's map of 1605. Floating trees described by Frobisher and shown by Resen suggest that the Gulf Stream was taking a northerly course, toward Iceland. Adapted from Resen map shown in Cumming, Skelton, and Quinn, 1971, p. 226.

the climate Normandy has now. In colder times, it had the climate southern Yorkshire has now (Lamb, 1966, p. 192). The difference is about 350 miles. In more important terms, Lamb says, the growing season changed by 15 or 20 percent between the warmest and the coldest times of the millennium. That is enough to affect almost any type of food production, especially crops highly adapted to use the full-season warm climatic periods.

Gulf Stream headings

Oxford, where the series of English readings comes from, is at the latitude of the Aleutian Islands, and lower Hudson's Bay in Canada. England's more southerly climate results from the Gulf Stream, which brings warm waters from the Caribbean north and eastward across the Atlantic. Let's look again at how the atmosphere, the ocean, and temperatures in England and western Europe are interconnected—and what each was doing in the little ice age.

If the westerlies around the North Pole expanded during the little ice age, they would frequently bring in air from the west, from the North Atlantic. As they expanded more, they would also often bring arctic air to England. Meanwhile, the same condition of the atmosphere, with westerlies expanded southward more often, would cool the North Atlantic. Those west breezes from the ocean would themselves become colder. And ocean currents would change also. Currents from the north could penetrate farther south, and the Gulf Stream would in turn be deflected to the south. It would not sweep by Iceland, as the earliest sea maps show, but would bend more to the south.

Gulf Stream records, though fewer and less detailed than records of wine harvests and glacial advances, confirm this. Earliest Gulf Stream maps show the North Atlantic around the time the little ice age began. The oldest map we have was drawn by Bishop Resen of Copenhagen in 1605 (figure 6.3). It draws on the voyages of the English explorer Martin Frobisher, and other sources of the late sixteenth century, and also, as Bishop Resen himself described it, on "an ancient map crudely drawn centuries ago by Icelanders."

Frobisher described the currents in considerable detail and told how driftwood "and even whole bodyes of trees" floated northeastward to Iceland (Stefansson, 1938, p. 54). This puts the Gulf Stream in a northerly position. But it was not to stay there for long. Though it's probably coincidence that Frobisher's three voyages found successively more ice, this clearly was a time of rapid climatic change. Later on, large amounts of driftwood are no longer mentioned. Perhaps the floating trees Frobisher saw indicate a change in flooding patterns in eastern North America that accompanied this time of climatic change.

Observations between 1577 and our own time reflect the little ice age's cooling, and the warming after three centuries or so. Among those who studied Atlantic currents were Ben Franklin, moved to action by the complaint about slow mail boats, and his nephew Jonathan Williams. Franklin made several Atlantic current maps, and took some temperature readings during his own ocean crossings (1786, 1834). Charles Blagden, an army physician, also studied the Gulf Stream, and published his results in 1781. The knowledge of old sea hands helped these investigators, as did instrumental

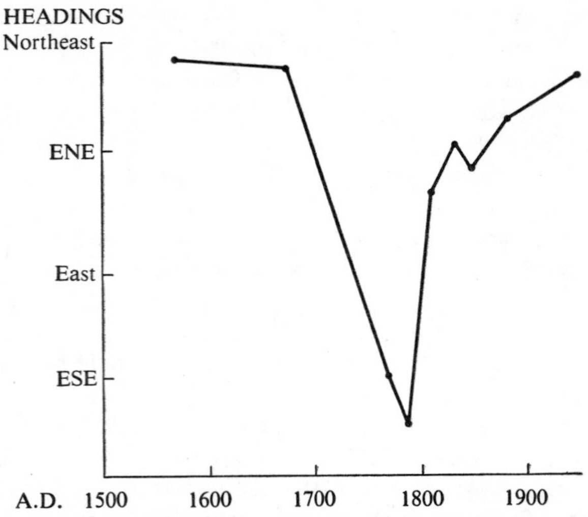

Figure 6.4. Chart of Gulf Stream headings, sixteenth century to twentieth century. Adapted from Stolle, 1975, p. 15.

findings–thermometer and wind readings. Efficient crossing of the Atlantic depended on an understanding of the currents, and, as commerce increased, this understanding became more important. The nineteenth century brought a number of bottle drift experiments that showed in more detail where the Gulf Stream flowed at that time (Lamb, 1972, p. 335).

At the University of Wisconsin-Madison, Hans Stolle (1975) collected references from many of the old maps and records and converted their information into one chart of the stream's position over the last 400 years. For this purpose, Stolle focused on the direction the stream took as it moved away from North America: its compass heading in this region is a key to its entire course. He developed information for the chart shown in figure 6.4.

The match between this chart and our other climate indicators from the little ice age is not perfect. The Gulf Stream seems to have shifted south more slowly, and moved back north more quickly, than we would have expected. Perhaps it is less affected by a cooling of the hemisphere than are some other indicators of climate, or perhaps the old records contain some misreadings. If we had hundreds of records instead of a dozen or so over those centuries, perhaps our picture would look different. But whatever the details, there is no mistaking the Gulf Stream's southward shift during the little ice age. Thus another source of evidence, completely independent of Alpine glaciers, wine harvest dates, and thermometer readings in England, tells of the same climatic change.

Meanwhile, in the States

To this point, our discussion of the little ice age has covered central and western Europe, Iceland, and the North Atlantic. For the United States, there are some pollen studies for the period which offer evidence of a little ice age here too. One study from northern Wisconsin shows "cooler and/or wetter conditions [beginning] about 600 years ago and continuing to about 100 years ago" (Swain, forthcoming). A preliminary temperature calibration of that record indicates a variation of about 1°C (1.8°F) between various intervals in the past 2,000 years.

There are fewer written records of climate on this side of the Atlantic during the early days of settlement, but we do have some, especially for the nineteenth century. And much of what we have is easy to interpret, because it came from weather instruments.

Since the founding of the United States Weather Bureau (now the United States Weather Service) in 1872, we have weather records with these characteristics:

1. They come from many stations, distributed throughout the country. (In western states, coverage developed somewhat later than in the East.)
2. The records are continuous. This is important, because if a station stops operating for a week or month, someone has to make an estimate for that month to complete all the averages—the yearly average, the average for the decade, and so on. (But it is possible to do that without much error if the gaps are small.)
3. The records are taken with standardized instruments and procedures. It is essential that thermometers and barometers give accurate readings and that methods be consistent; some thermometers in the sun and some in the shade add up to confusion.

Records predating 1872 also exist, though they are not so satisfactory. But they can tell us a good deal about weather and climate. It is important to get out of them all they contain.

Table 6.1
Average Monthly Temperatures for Portage, Wisconsin
1931–48 and 1829–42

	Portage 1931–48	Fort Winnebago 1829–42	Degrees difference in 1800s
January	20.9°F	19.2°F	-1.7°F
February	21.9	18.1	-3.8
March	32.7	33.8	+1.1.
April	47.5	46.1	-1.4
May	59.6	56.9	-2.7
June	69.5	65.9	-3.6
July	74.6	70.9	-3.7
August	72.4	67.3	-5.1
September	64.3	57.4	-6.9
October	52.6	48.1	-4.5
November	37.3	32.3	-5.0
December	24.5	20.9	-3.6

Source: Wahl, 1968, p. 74.

Professor Eberhard Wahl and others at the University of Wisconsin-Madison have analyzed many pre-1872 United States weather records and summarized the information in maps (Wahl, 1968; Wahl and Lawson, 1970). Part of their work was to determine how much faith to place in those records. They found that some eastern United States temperature records date from the early eighteenth century. For the early nineteenth century there is a scattering of records throughout the country, and from 1850 on fairly good general coverage. Who was taking these readings?

In the 1870s the Smithsonian Institution described 212 stations that had recorded weather in New York. Sixty-three were denoted academy, college, or such; another 14 were military (Wahl, 1968, p. 74). These people could tell the direction of the wind, and could read a thermometer and calibrate it periodically with freezing and boiling water.

Army records are especially important in the West, in Indian country. At first the surgeons at each post had the task of recording the temperature, wind, and precipitation; a handbook gave them thorough instructions. Later the Signal Corps, then the Weather Bureau, took readings.

Wahl's analysis of the records of Fort Winnebago, near Portage, Wisconsin, is an example of what we can learn from early instrument readings. Overlooking one of the traditional portage trails of the early French voyageurs, fur-traders, and explorers who made their way from Montreal to the Mississippi, Fort Winnebago kept weather records that begin in 1828 and continue, with some interruptions, until 1845.

Wahl compared the temperatures recorded then with records from 100 years later--1931-48--at the United States Weather Bureau station in Portage. Table 6.1 is one chart of his findings.

In every month except March, the 1800s were cooler. The biggest differences were in autumn—almost seven degrees in September.

The month-by-month comparisons are more important than the yearly averages because they have greater implications for food production. We saw that wine harvest dates don't respond to winter temperature changes, and the same is true for other agriculture. Within certain limits, summer temperatures are not critical either, for food. But spring and fall temperatures determine the length of the growing season. A small drop in average temperature makes all the difference in September, when some crops are still green and the nights cool to perhaps -2°C (+28°F) instead of +2°C (+35°F).

How certain are we of Table 6.1's differences between the 1800s and the 1900s? Although the Fort Winnebago temperature records show frequent calibration of the instruments and careful attention to details, Wahl looked for a way to confirm them.

Wind records were the answer. No matter how little faith we might place in the Fort Winnebago thermometers, those weathermen surely knew which way was north. They recorded the wind direction, and the records of 1829-42, compared with 1931-48, show that in every month of the year in the earlier century there were more northerly winds. For instance, in Januaries recorded at Fort Winnebago, the wind came from points between northwest and east 56 percent of the time. In the twentieth century, Portage records show that the wind came from these directions only 34 percent of the time.

For Septembers, the figures were 47 percent and 27 percent. Around the 1830s, then, September winds came from the north almost twice as often as during the 1930s and 1940s. More north wind would seem to confirm cooler times in the 1830s. Wahl made a detailed analysis that compared month-by-month the winds with the temperatures. He found they correlated very well. That is, knowing the winds, he could predict what the temperature differences would be. It's clear the 1830s were cooler.

Do these cool readings from Wisconsin agree with the evidence for expanded westerlies and a "little ice age" in Europe? An expanded system of westerlies means an expansion of arctic air, which should have a more frequent and stronger influence on an area like Wisconsin that is always under arctic air part of the year. An expansion of the westerlies greater and more persistent than the expansion of Mill Creek times would bring cooler weather to Wisconsin.

As in our studies of Mycenae and Mill Creek, a record from one location isn't enough to confirm a climatic change. Wahl considered

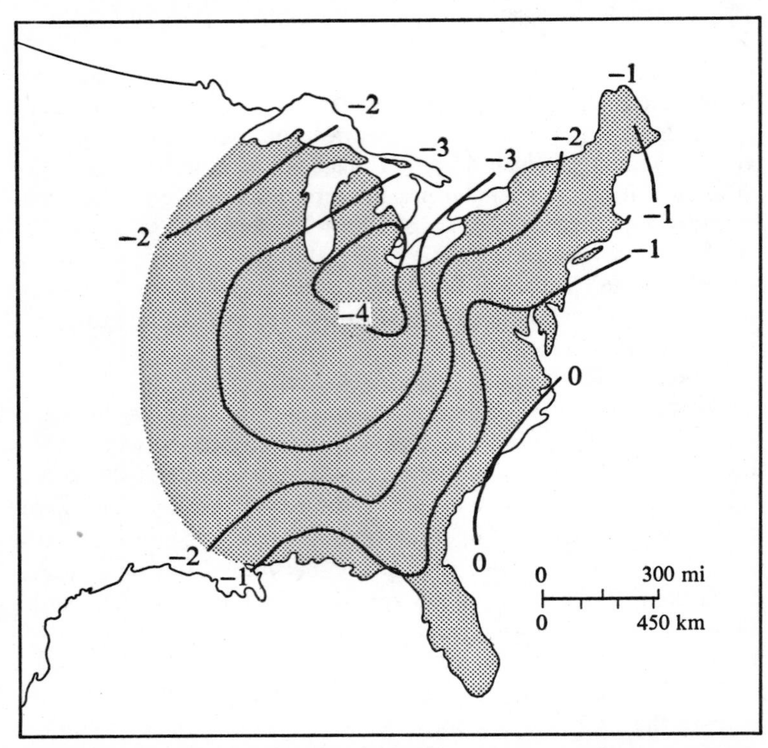

Figure 6.5. Early fall temperatures for the eastern United States: comparison of the 1830s with 1931–60. Numbers indicate, in °F, how much colder the falls were in the nineteenth century. After Wahl, 1968, p. 77.

much more than Fort Winnebago. He used 221 series of temperature records from the eastern United States to prepare maps of 1830s temperatures, as compared to the 1931-60 period. His maps describe various times of the year. The early fall map (figure 6.5) shows

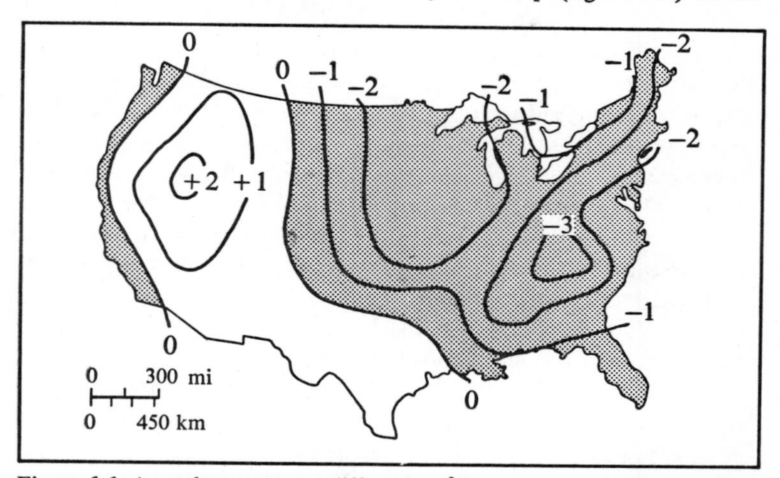

Figure 6.6. Annual temperature differences (°F) between the mid-nineteenth century and 1931–60 in the United States. Areas cooler in the nineteenth century are shaded. After Wahl and Lawson, 1970, p. 260.

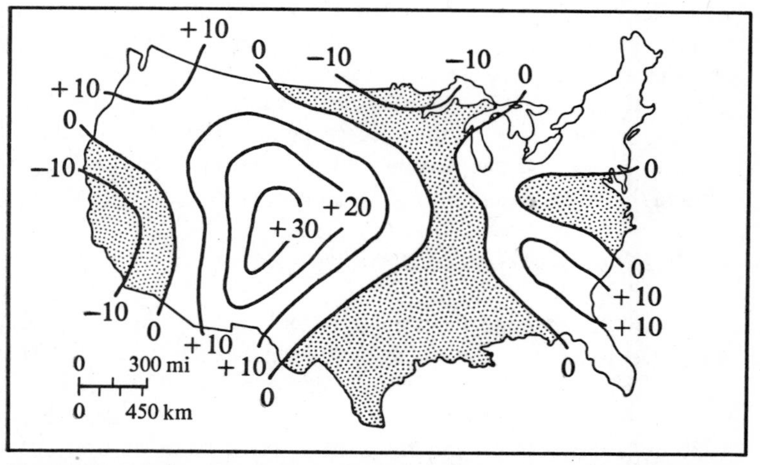

Figure 6.7. July-August rainfall in the United States: comparison of the 1850s and 1860s with 1931–60. Numbers indicate percentage changes. Areas drier in the 1800s are shaded. After Wahl and Lawson, 1970, p. 260.

greatest changes, although some extreme changes like those at Fort Winnebago/Portage are not apparent because the maps describe general patterns.

As people moved west in the mid-nineteenth century, they recorded weather information in new areas. Still, information is sketchy for parts of the West, even after founding of the Weather Bureau in 1872. To make national weather maps for the 1850s and 1860s, Wahl divided the country into 35 regions for which he had at least two sets of records–generally far more. Figures 6.6 and 6.7 are two of those maps, annual temperature, and July-August rainfall.

Wahl found generally cooler temperatures over the United States in the mid-nineteenth century; the exception was the Rocky Mountain area and the Southwest. The Rockies often experience a climatic change as if they were on the other end of a teeter-totter from the rest of the country and the hemisphere as a whole. They can turn warm while other regions cool, and vice versa. The Rockies are a big enough physical barrier to affect the flow of the Westerlies, and they generally "anchor" one of the northward loops in the westerlies' pattern. Meanwhile the midwestern and eastern United States are far enough downstream to be influenced differently by changing westerlies.

We have seen that in the 300 years from about 1550 to 1850 the westerlies expanded, bringing generally cooler temperatures and patterns of rainfall different from those before or since. The average temperature changes were small–generally a degree or two Fahrenheit. Yet, the effects were not small. Some of the winters were terrible, and not only in the coldest regions. The vintner suffered, and the farmer, and people went hungry. The slight shift in westerlies and the temperatures moved the lives of people and nations.

1850–1950: The climate turns "normal"

People view climate the way they look at the world generally: their recent experience is apt to be the standard, the yardstick. This is how we have measured the climate of the past 1,000 years, too. In the thirteenth century Iowa was dry (compared to our day). From 1550 to 1850 Europe and the United States were generally cold (that is, colder than they are now). That was a little ice age, compared to what we know. People don't say the 1920s were part of a little tropical age, compared to the standard of 100 years before.

We should not lose sight of the implications of this self-centered view, which can make us think that climate always was, or at least is supposed to be, what it is now.

The common use of "normal" weather readings reinforces that opinion. What are called "normal" readings are, by international agreement among weather agencies, 30-year averages. The "normal" temperature on July 4 is the average July 4 temperature of a recent 30-year period. The "normal" May rainfall is the average of what has fallen each of those 30 Mays.

The period that had "normal" weather is changed every ten years. A few years ago our normal weather was that of 1931–60; it is now that of 1941–70. This lockstep shift of the reference period does more than suggest that the last human generation or so of climate is all that matters. It smears out climate trends that develop over a period of several decades. For instance, if temperatures have persistently dropped since 1950, current readings would look nearly "normal" since by definition "normal" is largely based on a cooling period.

We've already covered evidence for the warming of the hemisphere in roughly the years 1850–1950. We have done so indirectly; this is the other side of the coin which showed the little ice age to be cold. Looking over the records, a good guess appears to be that average temperatures increased about 0.6°–1.1°C (1°–2°F) in those 100 years, throughout the latitudes of Europe and the United States. The rise was not steady or evenly distributed; in some places it affected mainly one season and in other places another. The warming trend reversed in the 1880s; the Krakatoa volcano exploded in Indonesia in 1883 and threw a great veil of dust around the hemisphere; perhaps this and other volcanic eruptions of that decade brought the several cool years.

But the trend is persistent and obvious. Some examples:

1. Gordon Manley found that central England temperatures rose throughout the 100 years or so after 1850. At first winters warmed up, and later springs and summers did. The period 1925–54 was 0.9°C (1.6°F) warmer even than the previous 25 years (Ladurie, 1971, p. 82).

2. H. H. Lamb found that the average length of the growing season at Oxford in the 1930s and 1940s was two or three weeks longer than in the nineteenth century (Lamb, 1966, p. 180).
3. The annual average temperature in Copenhagen rose about 1.4°C (2.5°F) during the 100 years (Ladurie, 1971, p. 84).
4. The glaciers in Europe retreated well beyond their village-crushing advances of the little ice age. The tax records and other documents uncovered by Le Roy Ladurie show this. Even more striking are old engravings, and photographs from the nineteenth century, which confirm the positions of the glaciers then and the changes since (Ladurie, 1971, between pp. 144 and 145).
5. Wahl's work with hundreds of past United States records shows that this country, except for the West, had consistently cooler temperatures in the early and mid-nineteenth century than in this century.
6. In Iceland, in the sensitive North Atlantic climatic border zone, mean temperatures rose about 1.4°C (2.5°F) between the late nineteenth century and the mid-twentieth century. They started their upward swing there later than in other places; a dip in the late nineteenth century could have come from the dust of the Krakatoa volcano and others (Bergthórsson, 1969, p. 98).

In this century, as generally over the last 1,000 years, Páll Bergthórsson's description of climate meshes with the instrument records, glacial advances, wine harvest dates, and other natural and human records of conditions in Europe and North America. We don't yet know enough about climate to understand every month or every year in terms of the general patterns we perceive, nor do we understand all the fine details of climate variations from place to place. But in considering what our climate *can* be, the best instruction is the past: what it has been. It is clear that the early and middle years of our own century—the time on which we base our ideas of normal climate—have been quite different from the 300 years that came before. We have had one of the least normal climate patterns of the last 1,000 years, *not* the only pattern. In 1914, Otto Pettersson stated:

> In the last centuries of the middle ages a series of political and economic catastrophes occurred all over the then known world. They synchronise with occurrences of a startling and unusual kind in

> the kingdom of Nature. The coasts of Iceland and Greenland became blocked by Polar ice. Frequent volcanic eruptions occurred in Iceland and the surrounding seas. Violent storm-floods devastated the coast of the North Sea and Baltic. In certain cold winters Oresund and the Baltic were frozen over and the lucrative Hanseatic herring fishery of the early middle ages which had been carried on in the Baltic and Oresund ceased altogether.
>
> All these events are recorded in ancient chronicles which also depict the social and economic state of the communities, which were greatly influenced by these violent climatic variations and their consequences: famine and disease. . . .
>
> Till quite recently the opinion has prevailed among meteorologists and geographers that the old records are unreliable and exaggerated and that no real variation in the climate has occurred in historic time. . . .
>
> Of late, however, dissenting opinions have been advanced in various countries.

It is now some 60 years after Pettersson wrote, and the illusion that climate is stable and unchanging is finally becoming untenable. We turn now to these, our own times.

III THE MONSOONS FAIL

CHAPTER 7

Death in the Sahel

IN THE LATE 1960s the outside world began to notice the Sahara's desert climate creeping southward into the Sahel, the semi-dry region half the size of the United States that lies across six African nations (figure 7.1). Through 1973, the 20 million people there suffered a drought that destroyed their pasturelands and their grains, dried up their wells and rivers, killed over a third of their cattle and more than 100,000 of their brothers and sisters. Until a million tons of food came in, much of it from the United States, a United Nations official feared that six million people might die (Bryson, 1973, p. 367). The summer monsoon rains—the only rains that come to the Sahel—had failed.

Rains in 1974 and 1975 increased the Sahelian harvest. But many of the nomads taking their herds back north toward the Sahara found dead trees and earth without grass or topsoil, littered with the bleached bones of cattle.

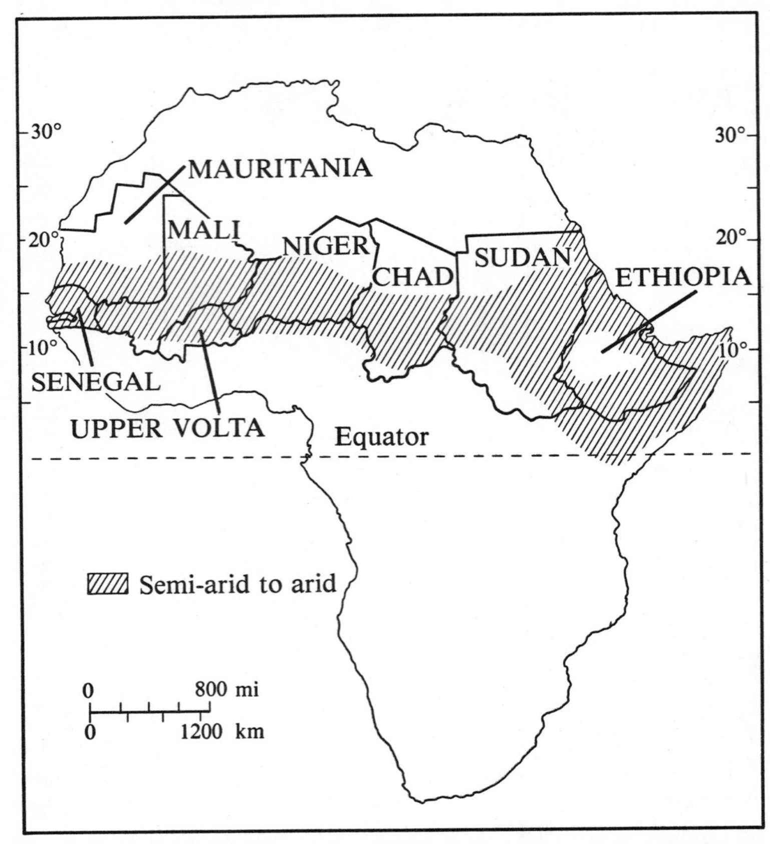

Figure 7.1. The semi-arid to arid zone on the southern edge of the Sahara desert is the Sahel. The Sahelian countries are those west of Sudan, but the semi-arid region extends to the eastern coast of the continent. Adapted from Bryson, 1973, p. 367.

Wind and rain had eroded the earth; sand had drifted into the wells. Where the rain and the soil did exist for grains, the increase in fuel and chemical prices prohibited the poverty-stricken countries in the Sahel from raising the food they might have. And for the millions still in refugee camps, crowded together with no work, little medical care, and poor sanitation, there is only the hope that Allah will provide a better life next year—or later.

Hopes and problems

The drought hit very hard in this band of poor countries that lies across Africa, from Ethiopia on the east through Sudan and the Sahelian nations: Niger, Senegal, Mauritania, Mali, Upper Volta, and Chad. Consider Niger, nearly twice the size of Texas with four million people (Du Bois, 1974). Most live in the far south, the only region of more or less dependable monsoon rains. To the north are rangelands which give way to dry country with a few partly green valleys and oases. Beyond is the Sahara.

People in the north are nomads, mainly Tuareg. As drought moved down and began to kill off their cattle, sheep, and goats, the nomads who survived moved south too and gathered in refugee camps near the cities. For camps on the fringe of drought, large rivers provided some water. But they became stagnant and too often contaminated by human sewage. Epidemics broke out, and Niger in 1973 had only 81 doctors. Few refugees could find jobs. The economy had little to offer; in 1970 Niger's per capita gross national product was $82. The total 1972 government budget was $47 million (Du Bois, 1975, p. 6). Since most nomads are Tuareg and the citydwellers are Songhai, in some places tribal and racial feelings festered.

Niger has no sea coast; the capital, Niamey, is 640 miles from the nearest ocean port. This has been a great problem for bringing relief supplies, and is a barrier against development of the country, which has only 300 miles of paved roads (mostly one-lane) and no railroads. Relief supplies came in by airplane, but that cost three or four times the value of the grain delivered. The president of Niger said that the transportation costs would have paid for irrigation systems along the Niger River that could supply as much food or more. But time was short, and would the river serve that purpose in a long and severe drought anyway?

The problems have been much the same in the other nations of the Sahel (all of them former French colonies). Politics has aggravated the situation; in Niger and Chad military coups overthrew presidents.

To the east, in Ethiopia and the nations near it, the drought may have taken an even greater toll than the estimated 100,000 deaths in the Sahel. A civil war in Ethiopia that brought the overthrow of

Emperor Haile Selassie was at least in part a result of the drought. In the east, however, many problems were concealed, so not all the consequences are known. As a peace corps director in Africa said, "The relief averted mass starvation. Getting the people back even to their fragile way of life is something else" (Mulligan, 1975). It will, indeed, take years, according to those who know the Sahel–presuming good weather, that is. Since people believe the drought is "broken," resettlement and rehabilitation are under way.

Hope in Allah, in relief supplies, in agricultural technology, and above all in good weather may be well placed–or may not.

The monsoon rains sweep from the South Atlantic and across the southern coast of Africa's northwestern limb, watering the row of coastal nations, then diminishing toward the Sahara. Since the mid-1950s, these monsoons have stayed farther south than before. Will the next decade or two be good, as the years 1974 and 1975 were, or will they be more like the six years before that? No one knows, but with 20 million lives in the balance in West Africa alone, it is well to consider the possibilities.

In India and other parts of southeastern Asia, even more people depend on the monsoons. India is about one-third the size of the United States; it has 600 million people, as opposed to the 200 million in the United States. Seventy percent of India's rain comes from the summer monsoon, and some monsoon failures there have reduced grain production by one-third (Das, 1968, p. 32).

In both Africa and India, the sections most likely to experience drought are at the northern edge of the monsoon's range, far from the sea. These, like several other areas we have looked at, are climatic borderlands–ecotones. They are the most sensitive to climatic change and have the most variable weather. In fact, the name *Sahel* is Arabic for "edge," the edge between the Sahara and the more moist lands to the south.

In India we will, in chapter 8, also focus on another kind of area subject to monsoon failure: an area quite near the sea and one which lies under the moist monsoon winds, and yet peculiarly is desert.

Sea breezes and monsoons

The monsoons that mean life or death to hundreds of millions are not, of course, isolated phenomena of the atmosphere. They fit into the pattern that includes the sweep of westerlies around the poles. We have seen the importance of the westerlies to Europe and North America, but most the world's population lives elsewhere. The monsoons also have ties to the westerlies.

Monsoon also comes from an Arabic word, meaning "season." The monsoons are large-scale wind systems of the tropics that bring alternate wet and dry times each year. The Asiatic monsoon is the best known, but the African is much like it, as is the rainy season of northern Mexico, and those of northern Australia and a few other areas.

Several major forces of the earth's weather systems join together to produce monsoons. Therefore the monsoons are regular, permanent features of the atmosphere. But they do have beginning and ending times each year, and they have geographic limits. As with any other feature of climate, the borders in time and space are variable, and such variation often makes the difference between living and dying, as in the Sahel.

The monsoons are related to the flow of westerlies, the rotation of the earth, the seasonal changes in heat received in various parts of the earth, and the different ways sunlight affects land and sea. These are governed by physical laws, and although the interactions are complex, the parts mesh beautifully, like a fine machine or, rather, like a system of nature. The world of air is one world, as yet imperfectly understood. But within the past few decades meteorologists have come to see the overall features.

One important feature, known centuries ago, is wind caused by uneven heating of land and sea.

The sun is a driving force behind all movement of our atmosphere. It causes convection currents, like the currents in a pot of boiling water. To picture how such currents move, think of a boiling spaghetti pot, with the spaghetti showing the water's movement. The water currents swell upward in the center, where heat is most intense. They shoot outward and descend along the relatively cool sides of the pan. So it is, roughly speaking, with coastal breezes, monsoons, and other large currents in the atmosphere.

Along the coastline on a sunny day, ocean and land are warmed by the sun, but each differently. The land surface, especially if it has little vegetation, warms rapidly. Most of the heat stays concentrated

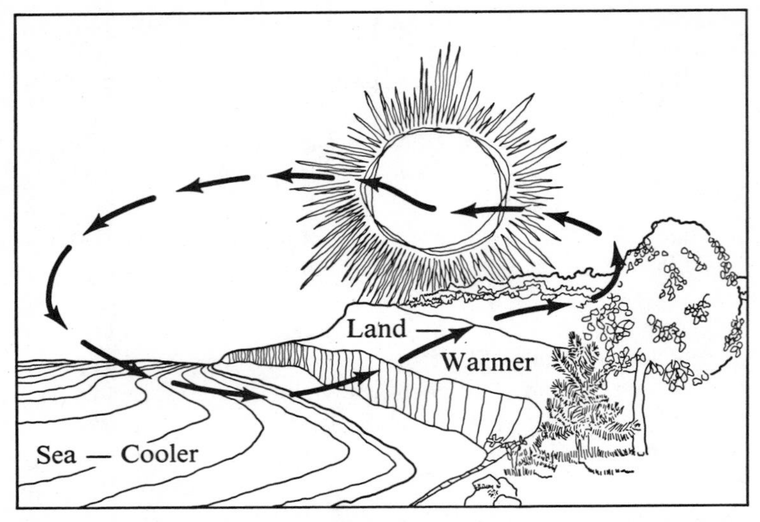

Figure 7.2. Daytime sea breeze. The circulation, of course, extends farther over the sea and over the land than shown here.

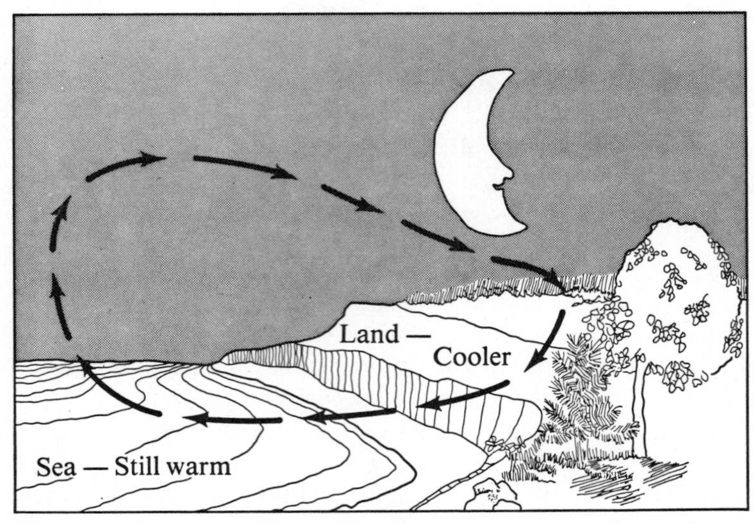

Figure 7.3. Nighttime land breeze. The circulation extends farther than shown.

near the surface; little penetrates downward through the soil. So the temperature of the land surface rises, as does that of the air above it.

In the ocean, sunlight can penetrate downward many feet, warming to some depth immediately. As the water mixes, so the sun's energy is spread through a still greater mass. Water requires more heat than does soil to warm up the same number of degrees, so the ocean surface, and the air above it, stay relatively cool.

The warm air above the land rises like water in the center of a spaghetti pot. Some distance above the earth it flows out over the sea. Within a continuing cycle it sinks, replacing the air below it, which moves to the land to replace the air that rose (figure 7.2).

If you stand on a sunny beach you cannot see this whole cycle. But you can feel the hot sand on your feet, water cool to your touch, and the wind in your face as you look out to sea.

This is the daytime sea breeze that helps make coastlines pleasant on summer days, and can also bring showers. As the moist ocean air rises it expands and cools, and moisture condenses.

At night, the whole area cools off, loses heat, and the land cools off faster, just as it warmed up faster. So the winds reverse. The cool air over the land sinks and moves out to sea, at the surface, where it rises and flows back over the land (figure 7.3).

These patterns can be overruled by a stronger weather system, but are themselves strong and persistent.

Monsoon winds

In the tropics, another type of regular, persistent wind blows—a larger-scale wind that arises from the same type of conditions that produce land and sea breezes. That is, it results from a cycle of heated air rising, then being cooled and sinking. But in this case land-sea temperature differences are not the whole cause. Because of the earth's spherical shape, the sun heats the lower atmosphere in the tropics more than to the north or south. Air in the tropics rises, and part flows north, part south. After traveling some distance it sinks, and a flow moves back toward the equator, from the north and from the south.

The equator is usually not the center of this pattern. In July, when the sun's heat is most intense north of the equator, the region of rising air is also north of the equator. In January, the center of

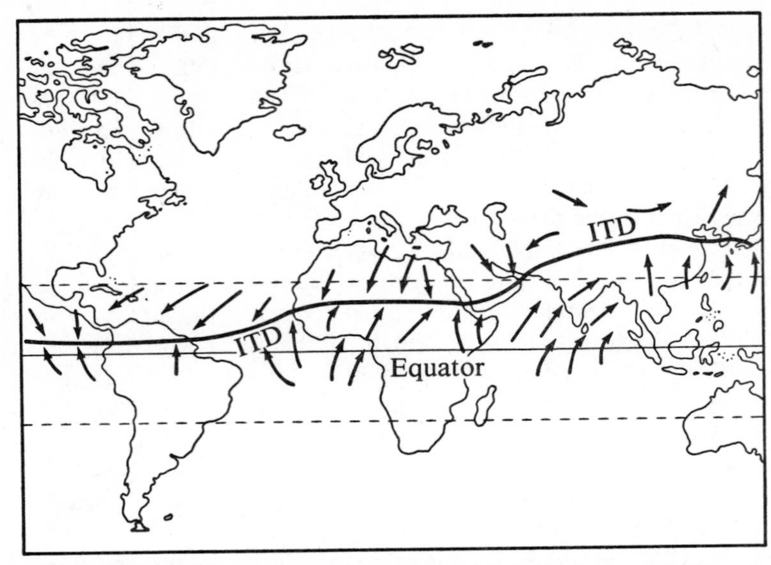

Figure 7.4. Typical surface wind patterns and the ITD–intertropical discontinuity–in July. Adapted from Das, 1968, p. 3.

the pattern is farther south. The region of rising air is not straight or even. It is positioned in part by the fact that land and sea heat up differently under the sun's rays. So the zone where winds come toward each other and rise moves north and south with the seasons. It is called the ITD, the intropical discontinuity.

The ITD is north of the equator in July; one possible location is shown in figure 7.4. Note that in many places the surface winds flow toward it from both north and south, as we have described. Then they rise, although neither the upward movement nor the flow of air at high altitude shows on our map.

In many parts of the tropics, the seasonal movement of the ITD changes the direction of the winds and the source of the air. Monsoons are one form of these seasonal wind patterns. The sun's energy, the tilt of the earth's axis that brings the sun directly over a given area at a certain time, and the distribution of land and sea all help make the monsoons blow when and where they do.

The monsoons are most important in those parts of the world that depend on them for rain, and that happens in only a few areas—though large and populous areas—like the Sahel and India.

Rainy summer monsoons come only to land between sea and the ITD. In such a place, the wind blows from the sea, over the land and to the ITD. As it does so, it rises over the warmer land, as would a simple sea breeze, and its moisture condenses out as rain. As the map shows, India and northwestern Africa, in the Northern Hemisphere, meet these conditions.

In winter, the Indian monsoon blows from the north or northeast—from land to sea, rather like the nighttime coastal breeze. Winter is analogous to night, a time of cooling off from the temperatures of summer. The ITD has moved south, over the ocean. Winter monsoons are generally dry winds, having no source of moisture. An exception is part of the winter Indian monsoon, which sweeps across the Bay of Bengal and brings rain to the southern tip of the subcontinent.

While the monsoons are a large-scale spaghetti-pot circulation that takes place in the tropics, this type of circulation influences weather away from those regions too. But it can *dominate* in the tropics because they are outside the flow of the westerlies.

For the Sahel, India, and other monsoon lands, the critical question is where (and when) the rain falls. Often, the problem is monsoon rains that do not reach far enough north—this happened recently in the Sahel (figure 7.5). The position of the ITD is critical: monsoons cannot penetrate past it.

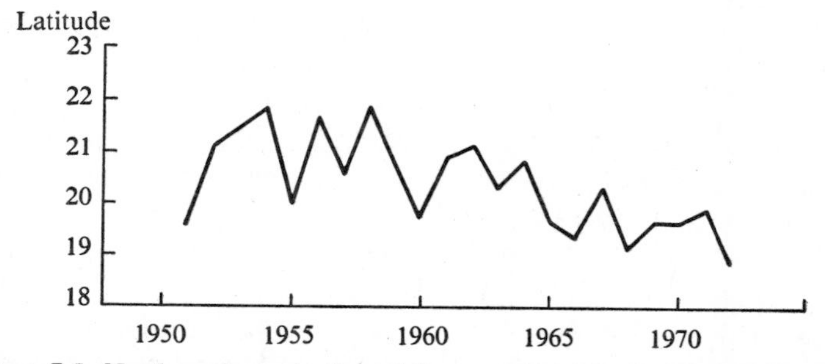

Figure 7.5. Northward penetration of the monsoon rains in North Africa (in the Sahel). The higher the latitude to which the rains penetrate, the greater the seasonal rainfall in the Sahel. Drought prevailed in the region from 1968 through 1973. After Bryson, 1974b.

Why the rains stay south

Even in tropical Africa, the westerlies that sweep around the North Pole influence climate. In the most simple terms, westerlies that are farther south will hold the monsoon rains south. They correspond to less rain in northern India, and less rain in the Sahel—an effect called by Bryson (1973) the Sahelian Effect.

Any climatic change comes to different places in different ways. Rainfall in the Sahel will not change in lockstep with that in northern India. But the two are linked. In 1972, with the Sahel drought at its worst, parts of India had a 60 percent shortfall in harvest. Bangladesh, at India's northeastern tip, had a rice harvest 2.5 million tons below that expected. Chinese newspapers spoke openly of drought and famine (Bryson, 1973, pp. 367–368).

In earlier times also, the widely separated monsoon lands have suffered drought together. About 1500 a great Mali empire in West Africa collapsed. In northern India, the emperor Akbar began to build his new capital city of Fatehpur Sikri in 1570. Fifteen years later Fatehpur Sikri, its water supply inadequate, was abandoned. During that same century the westerlies expanded greatly, bringing the little ice age to regions farther north, to Iceland and Europe.

A recent analysis at the University of Wisconsin-Madison shows that Icelandic temperature records correlate well with monsoon rainfall in northwestern India. Historically recorded famines there in turn coincide with low rainfall that shows up in this comparison. Famine often comes to India when Iceland is both cold and in the process of cooling. The fourteenth century and the nineteenth century contained a number of famine years in India, and in Iceland there were unsettled times with both warming and cooling, and with some very cold decades (Brinkmann, 1975).

To show more clearly how the westerlies affect the monsoons, we'll look at the atmosphere in a little more detail.

Picture again the ITD across the African Sahel in summer. Monsoon rains from the south come toward it but do not penetrate beyond. On the other side, the north side, air flows south toward the ITD. In the ITD region, it rises, moves back poleward aloft, and then sinks as part of the continuing flow.

This circulation does not extend all the way from the ITD to the poles. It does extend, roughly speaking, to the westerlies. And

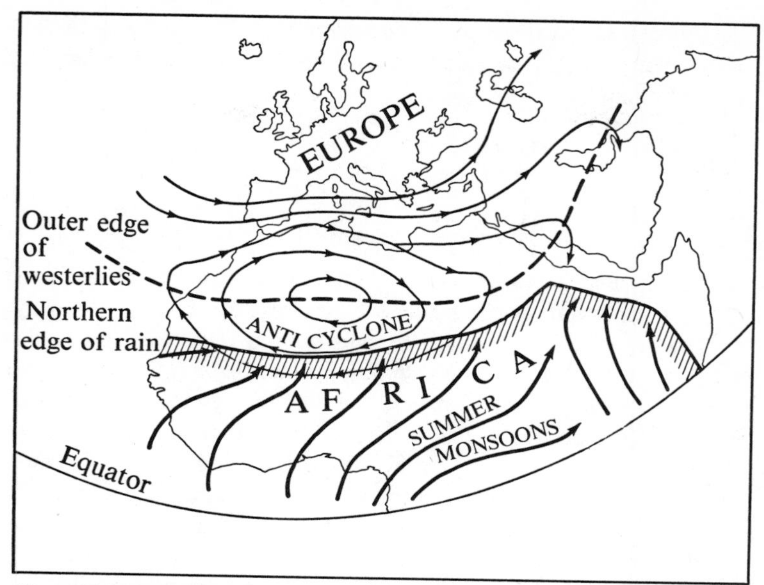

Figure 7.6. Simplified view of the relative positions of the westerlies, one of the subtropical anticyclones, and the ITD, in summer. The westerlies flow at high altitudes while the monsoon airflow is at the surface.

around the earth, at the latitudes where the air is sinking, and the outer edge of the westerlies is flowing by, is a series of large high-pressure areas–anticyclones. Such highs are areas of clockwise circulation, as is apparent on any weather map.

That motion fits in with the westerlies, which lie just to the north. As figure 7.6 shows, the eastward flow at the northern edge of an anticyclone is the eastward flow at the southern edge of the westerlies.

Areas of rising air and sinking air, of rain forest and of desert, are also in step with the movements we just described. You have seen that rising air expands and cools, and if the air is moist, rain can condense out. So rising ocean air produces monsoon rains.

On the other hand, sinking air, rather than giving up moisture, can take on moisture. It brings dryness, not rain. The air that sinks into the subtropical anticyclones is dry, and beneath them lie the world's great deserts, including the Sahara.

The crucial question for the Sahel and for northern India, then, is where the border between moist and dry air lies–indeed, where the whole system lies:

westerlies	north
anticyclone deserts	↑
ITD	↓
moist monsoon air	south

If the whole system moves north in the summer, the moist monsoon air goes north and people in those regions prosper.

The movement of the system, and the importance of the westerlies, is most dramatically seen in India. Along that country's northern border lie the Himalayas, so high they penetrate more than one-half the amosphere. The westerlies flow either to the north or to the south of this barrier. In winter the westerlies are of course expanded, and they flow south of the mountains. When–and if–they make the summer shift to the north, the tropical weather system moves north, and India gets monsoon rains.

If this movement comes every summer, people of the monsoon lands prosper. If the system stays farther south for most or all of a couple summers, green fields become desert and thousands die. If the monsoons stay south for years, how many millions will die?

CHAPTER **8**

A Manmade Desert

IN THE INDUS RIVER region of eastern Pakistan and northwestern India, now also called the Rajputana desert, an early agricultural civilization arose 2,500 years before Christ. Two other river-based cultures of the ancient world, those of the Nile and the Tigris-Euphrates, are better known, but the civilization of the Indus valley covered more area than those two combined, and lasted for perhaps 1,000 years.

Two great cities dominated the region, Harappa and Mohenjo-daro. Those cities are notable for their brick fortresses and huge granaries; the fortress of Harappa was 200 yards wide and over a quarter of a mile long. The granary floor area was something over 9,000 square feet. The Harappans grew wheat, barley, melons, sesame, and dates, and they may have been the first people to cultivate cotton. They occupied both the floodplain and areas where irrigation was not possible, nor was it needed with the amount of rain they had. They carried on a far-reaching trade in metal, wood, ivory, and their agricultural produce.

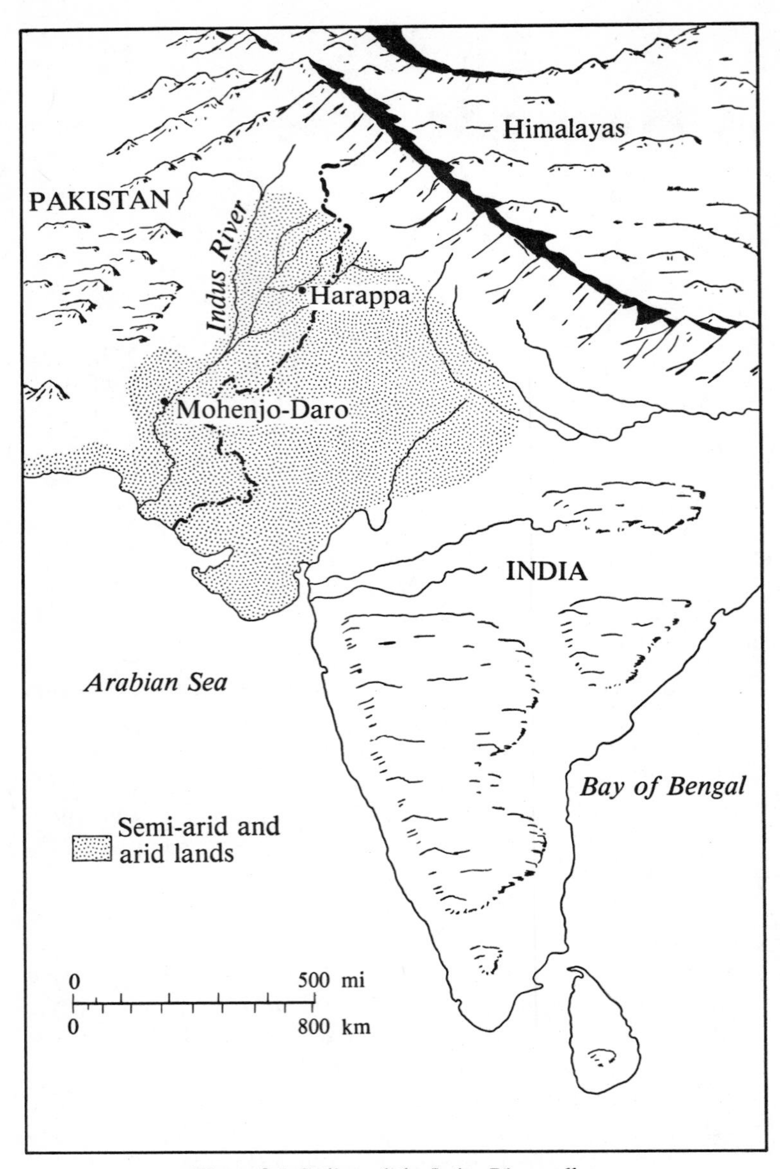

Figure 8.1. India and the Indus River valley.

Harappan cities resembled our own in some ways. They had grid-like plans with buildings of standard sizes and construction, and sewage systems. Many buildings were equipped with bathrooms, and earthware pipes encased in brick carried away the waste.

Beyond the broad outlines we know little about the Harappans. They made pictographs, but these have not yet been deciphered. The rise of the Indus civilization was, we know, rapid, and all the basics appear to be homegrown; there is little evidence of heritage from other advanced peoples of the day. Its decline, too, somewhere about 1700 B.C., was rather rapid; the reasons are not clear, though we do know that a changing climate contributed. And the Indus people themselves may have been partly responsible for that change. The present extent of the semi-arid and arid lands in that area is shown on figure 8.1.

Is the Harappan story another tale of monsoon failure? We believe that it is, but that in this case expanded westerlies are not the whole explanation. People, we think, altered the Indus valley climate, by accident. They helped make a dustbowl out of a breadbasket, and have kept it that way (Bryson and Baerreis, 1967).

Such desert-making by man may be partly to blame for drought in the Sahel and other monsoon lands today, along with changes in the westerlies. Humans can alter climate on a very large scale.

While Rajputana has never again been home for such a populous and organized civilization as the Harappans, other peoples have lived there. The next identifiable people were the Aryans who, some theories hold, conquered the Harappans. If so, they did it when not much was left to conquer. The Harappan ruins show signs of a slow decline and abandonment (Bryson, 1967, p. 53). In fact, radiocarbon dating shows a gap of something like 600 years between Harappan material and that of the Aryan culture (Wendland and Bryson, 1974). Moreover, in most cases the Aryans did not build their towns above the broad river valleys, away from the floodplains, as the Harappans did. They settled down within the valley, close to the river channel itself.

After several centuries in which the Aryans, too, appear to have declined, scattered settlements appeared again. Then, by about the fourth century A.D., the more extensive Rangmahal culture occupied the Indus region. Many dust storms were recorded in the seventh century, and by A.D. 1000 the desert had spread considerably again, and it has spread more in the recent past (Bryson, 1972).

Lakes fresh and salty

The remains of Harappan towns and cities do not themselves show why the culture declined and disappeared. But an Indian scientist, Gurdip Singh (1971), examined pollen in the beds of salt lakes in Rajputana, and reconstructed a sequence of climatic conditions paralleling the record of human settlement uncovered by archaeologists (figure 8.2).

He found that for thousands of years before 2000 B.C., the lakes held fresh water, and moist-climate vegetation grew. Then the lakes became salty, indicating less recharge with fresh water. Some dried up entirely a couple of centuries after 2000 B.C., about the time of the Harappan decline. Episodes of more moisture since then relate to times of higher populations. But the climate in which the Harappans prospered has never reappeared. The lakes have remained salty, and some have been intermittently dry. Plants (and animals) that grew where the Harappans lived now grow only on the fringes of the Rajputana desert.

For the time of the Harappan decline, then, field evidence showed decreased rain for northwestern India. We said in chapter 7 that drought there is related to Arctic cooling–to expansions of the region of arctic air. And the soils of northern Canada show that not only did the Arctic expand at the time of the Mill Creek drought (chapter 2), but an even bigger expansion occurred at the time of the Harappan drought (Bryson, Irving, and Larsen, 1965; Bryson, 1966).

We have sketched a few features in the history of the Indus valley over thousands of years. It is clear the climate changed from time to time. A worldwide climatic change may have helped make the desert and started the Harappans' decline. But the Indus region has remained dry, compared to 4,000 years ago, despite several subsequent global climatic changes. Some other forces must be at work.

Let us focus on the present, consider the conditions and the mechanisms now apparent, and see how they might have operated in the past.

The rain that doesn't fall

The Indus region today is dry, with a very dry area, the Thar desert, at the center. The Thar gets less than five inches of annual

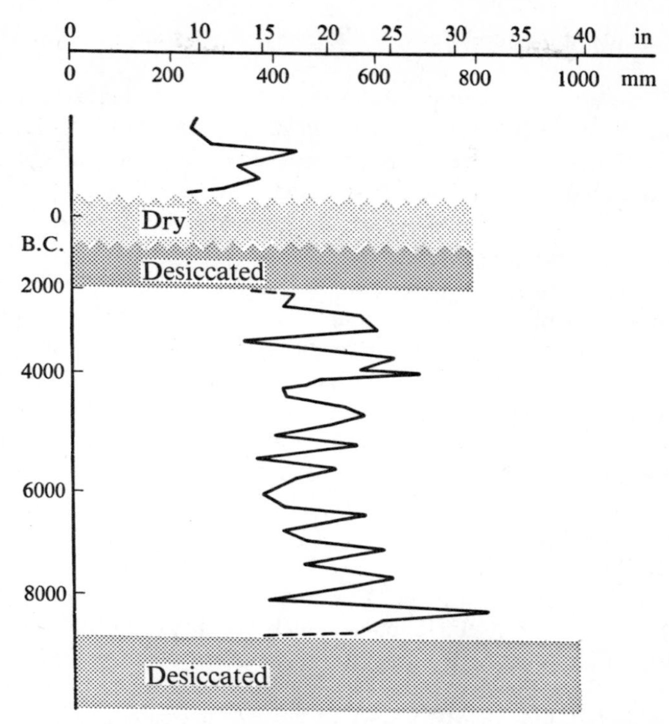

Figure 8.2. Amount of summer monsoon rainfall, calculated from pollen data, Lunkaransar, Rajasthan, India. Adapted from Bryson, 1975a; based on Singh, 1971.

rainfall. The surrounding arid region is a quarter of a million square miles or more, depending on the definition of "dry" used. That is 20 percent of India, and about the size of Texas. The perimeter of this vast desert, dotted with small villages and campsites, is advancing the equivalent of about half a mile a year. Away from the Thar itself some agriculture is practiced, but yields are poor, certainly not the harvests the Harappans knew.

And yet the air over Rajputana was and *is* moist. Summer monsoon winds from the Arabian Sea, moving over the desert, contain four times the water vapor of air over most deserts, and 80 percent as much as over tropical rain forests (Bryson, 1967, p. 53). These are clearly monsoon winds of the rainy season, and yet rain does not fall.

If the air were to rise, and therefore expand and cool, rain would condense out. But in Rajputana the air, far from rising, is sinking, absorbing moisture, creating the conditions we described for the subtropical anticyclones—a desert region.

Why does the air sink?

Indian scientists in the last couple of decades found this sinking—subsidence—over the Indus region. But they had difficulty learning the cause, for their calculations, which took into account the amount of heat coming from the sun, the carbon dioxide and water vapor in the air, and other factors, could account for only two-thirds of the sinking observed (Bryson, 1972, p. 139).

Climatologists at the University of Wisconsin-Madison investigated, and found something that, we believe, accounted for the difference: dust, which man may have started, and to which he certainly contributes today. Dust may be the seed by which the desert reproduces itself. Rajputana air is not just dusty the way an industrial city's is, or the way the midwestern countryside can be on a windy spring day. Over each square mile of Rajputana sometimes hangs more than five tons of suspended dust (Bryson, 1967, p. 53). That is several times the average density over Chicago: Rajputana air is very turbid indeed. During pollution episodes the air in large cities does become as turbid as the average air over the Indus area, but to lesser heights. The daytime Rajputana sun is a hazy red or is completely hidden. At night dust veils the stars.

The effects of dust in the atmosphere are complex and not fully understood. Dust reflects back some sunlight, but it also absorbs some, and in turn gives off heat. It has the same actions, in differing amounts, on heat coming up from the earth.

Work by Indian scientists and at Wisconsin indicates that because the dust over Rajputana keeps out some sunlight, the ground is a little cooler in daytime than it would otherwise be (Bryson, 1972, p. 141). This lack of warming means that not as many daytime updrafts which could bubble up moist air far enough to produce rain are created.

At night, the top of the dust cloud cools greatly, because the dust loses its heat to the coldness of space. As it cools, this air sinks. Also at night, the lower part of the dust cloud holds some heat to the ground, the way ordinary clouds can keep a fall night warm—but not warm enough for rising currents and rain. This warmth prevents the formation of dew that would hold down the dust a bit and water the grasses.

The most obvious source for all this desert-making dust is the desert itself. While that part of the world is generally dusty and dry, the dust is thickest over Rajputana, and the same kind of clay, montmorillonite, is prominent in the desert and in the air.

So the desert sustains itself, defying the moist air above.

How did the dust begin, and how does man contribute to it? We can envision an agricultural expansion in Harappan times: a growing ɔopulation, farming more land more intensively, moving out to over the whole region. After a time, as the global climatic change ccurred, some dry years came. The grass cover weakened and dust ·gan to blow into the air. As the dustbowl grew and times became ıgher, the Harappans tried to overcome their environment, work- the land even harder and raising more dust. The drought worsened. ally, much of the land had to be abandoned, and so did the cities. The abandonment perhaps gave Rajputana time to begin to re- ·r itself; the pollen records show that somewhat moister times ·wed. But whenever the vegetation began to improve, it became attractive and pressure on it increased, bringing desert back.

the air hasn't cleared

Rajputana today, the desert is growing, advancing into arable bout half a mile a year. Where grass starts to grow, cattle, or goats eat it. Forests around the desert have been cut for d and for other uses of India's increasing population. Where no wood, people use cow dung for fuel. The destruction of :over, the use of manure for fuel rather than to build the ·e too contribute to land destruction and soil movement. ›s if humans can make a desert they can unmake it. With so isture overhead, could the clouds be seeded? But the clay desert—the dust itself—makes excellent cloud-seeding nuclei, and yet rain doesn't fall. In fact, the clouds may be so overseeded as to discourage rain. With too many particles competing for water vapor, droplets can stay too small to fall as rain.

If some grass could grow back, there might be less dust and more rain, and the balance might tip away from desert. A fence built around a large field in the desert brought abundant wild grasses in two years, without planting or irrigation, simply by keeping out the men and goats (Bryson, 1967, p. 55). As similar experiments in the western United States show, grasses do grow if grazing pressure is lifted.

The problem is the continuing pressure from man. Suggestions that India entirely fence off tens of thousands of acres are impractical, for what would happen to the people who scratch out a life there now? A better experiment would be to build fences but allow cutting of grass for animals kept in pens. This would protect plant roots and shoots, which are trampled or eaten by grazing animals. If a large enough area could be fenced off to reduce the dust and bring back the rains, many thousands of square miles might become more like the way they were 4,000 years ago. But we don't know how many animals and humans Rajputana can support even in that way.

The nomads and farmers of Rajputana past and present (and of the Sahel too) are like us all. They did not intend to destroy their fragile land. They wanted a better life, or perhaps in their case, any life. On this earth today we burn fields in the tropics to plant crops, or we burn coal and oil, also wanting a better life—or life. In doing so we too make some deserts.

CHAPTER 9

The Enduring Problem

MILLIONS OF LIVES are at stake in the monsoon lands, and many theories have been offered to explain the reasons for and possibility of drought there. In the preceding chapters we have described how an expansion of the westerlies can hold the monsoons south, away from lands that need them. We have also suggested that regional dust, to which man contributes, can make desert where otherwise monsoon rains would come.

Other explanations have been offered. For instance, some emphasize the direct role of man in the monsoon lands. Overuse of the land can strip away every living plant. Erosion follows, by wind and by whatever rain falls. Topsoil blows and washes away. This process can destroy productivity even when rainfall does not decrease. Or perhaps the glaring desert sands, made more glaring by man and beast, reflect back sunlight the way dust in the air can. This would prevent the lower air from warming and result in less rain. Another argument is that the westerlies may be important, but that expanded westerlies, with much looping, can themselves reach areas like the Sahel at times, bringing some rain.

Still others say climate does not change at all in the ways we have described in this book, that it only shows random variation about a normal condition that lasts thousands of years. A farmer of the southern Sahel expressed a view like that after the rains of 1974. "We have turned a corner," he said. "This was the second drought I have lived through. I probably won't live long enough to experience the next one" (Gerster, 1975).

Which point of view is most accurate? If two good years signal the end of a drought, our problems are relatively small, in the short run. On the other hand, if a major climatic change can last many decades and can affect many parts of the world, as we have shown, even the best relief efforts will not help much.

Within the next few years we will learn more about climatic change. Meanwhile, whoever is right about the nature of monsoon failure and other changes, there is an underlying, enduring condition that will bring more suffering and death.

The problem is our increasing population and its demand for food. Clearly our planet cannot produce food for a limitless population; the human population cannot keep growing without end. In this discussion of climatic variation, we will consider briefly one aspect of the food-population problem: how does a *changing* food supply affect populations?

Our point is that population does not simply grow at one rate, food supply at another, slower rate. On the contrary, food supply can fluctuate quite rapidly, especially within a limited area, be that the Sahel, India, or some other part of the world. These sharp changes from year to year or over a few years bring suffering and death, and they tell something about the nature of our problems.

Recent research in Africa shows that the following sequence (see figure 9.1) occurs with cattle herds (Rapp, 1974). In pastureland that has more than enough grass for its cattle herds, the cattle population grows. After a time it reaches the carrying capacity of the pasture (1), but the herds are still growing. The pasture is overgrazed and begins to die out. As the cattle become weakened by hunger, diseases break out (2). The number of cattle drops rapidly, below the carrying capacity (3). Then the cycle can begin all over again, if there is no permanent ecological damage.

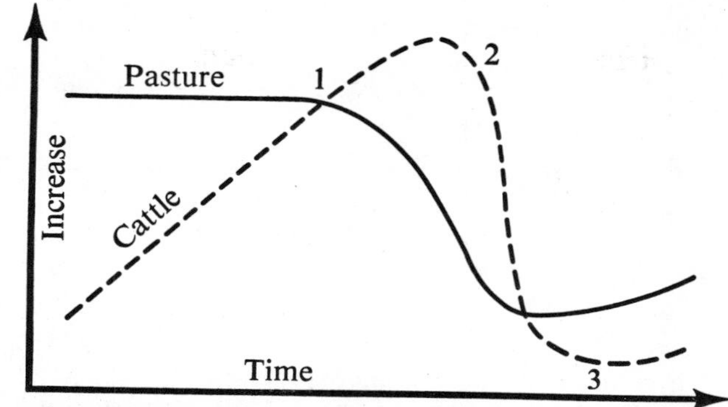

Figure 9.1. Observed effect of cattle overgrazing in Africa. (1) Carrying capacity reached; (2) starvation and disease cause many deaths; (3) herd below carrying capacity—cycle may begin again. After Rapp, 1974, p. 23.

Once the critical point is reached, then, food supply is not even maintained at a steady level that slowly becomes inadequate for an increasing population. The food supply itself decreases rapidly, leaving a huge gap between supply and demand.

Similar analyses might be done for human populations, but humans require a longer time for reproduction and normally live much longer than cattle. We suggest that the same pattern can hold true. And consider that the food supply can begin to drop even *before* the point of overgrazing or over-farming.

The following sequence, we believe, may have happened many times in human history. Picture again an area of rangeland, at a time when the amount of grass and the number of grazing animals are both relatively low. Suppose that because of a more favorable climate, or for some other reason, the grass supply in an area begins to increase. More food leads to more animals. As long as the grass supply stays ahead, the animals multiply. So do the humans who depend on the cattle. But when years of famine come, possibly because of a less favorable climate, the food supply drops rapidly. Too many animals and too many humans see their support fall away. They travel farther for the little food left. Cattle, sheep, and especially goats eat and trample the plants. Farmers dig more wells and take water from rivers, depleting the ground-water. They may plow larger fields, trying to get their small yield from more acres but only destroying ground cover.

Food stored or brought from the outside may carry the population peak onward even after local resources are declining. But as the need for food becomes critical, people die from starvation, from diseases related to malnutrition, and from poor sanitation. Epidemics can drop the populations very low, even below the small numbers the countryside could now support (figure 9.2).

In our age of airlifts and medical treatment, people don't usually die in such enormous numbers as they did 100 years ago. Perhaps in a given famine only tens of thousands will die, instead of hundreds of thousands, or hundreds of thousands instead of millions. As conditions improve in a land of famine (however long the bad times last) and the rangelands and fields are green again, the cycle of population also begins again. Another bust will surely follow the boom. Part of this cycle has been played out in the Sahel, among other places, in recent years. Given the opportunity, humans probably will proceed to build their populations up again in those places, perhaps no wiser for the experience than before.

The alternative to these times of mass starvation and death is to keep population near or below the number that can be supported in the *worst* of times, not in the best of times and not even in "average" times. To a certain extent stored food can support additional people, but it is impossible, or nearly so, to promise that

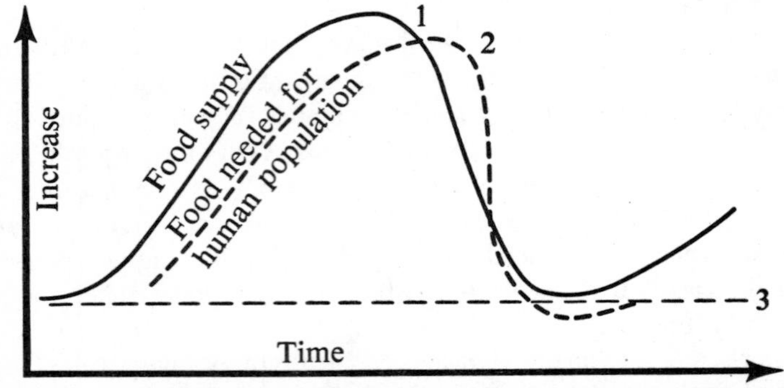

Figure 9.2. Projected results of human overpopulation associated with varying food supply. (1) Food supply and need are equal; (2) many deaths; (3) long-term carrying capacity.

enough food from good times will be held over for famines. The prospect of doing so should not be a rationalization for higher population. The pressure to sell food for profit in good times is great, stored food spoils or is lost, and no one can predict the time of onset, intensity, or length of the next famine.

Of course, efforts to increase agricultural production, to store and distribute food where needed most, and to make human lives more important than profit and power are worthwhile. But such measures may only increase the number of people who will starve to death while young if the population does not stay below the level of the *minimum* food supply that will be available.

To what area of land should this rule apply? It does not have to apply to every farm, or county, or every region within a country. But, first and beyond a doubt, the population of the earth as a whole cannot exceed the minimum food supply that will be available. Some people think we are already beyond that point, living on favorable climate, nonrenewable energy, and other good fortune that cannot continue (Ehrlich, 1974; Watt, 1974).

In reality, the self-sufficient units will have to be much smaller than earth-size.

No one knows what our food supplies will be. But it is vital to consider, as best we can, the possibilities. Surely food supply will be affected by climate. We have seen that several possibilities exist, other than a mere continuation of today's "normal" climate. And the worldwide linkage of climatic change can produce worldwide repercussions.

In the rest of this book, we will take a longer view of climatic history, to see what some other possible climates are. We'll also describe some causes of climatic change.

IV A PERSPECTIVE ON CLIMATIC CHANGE

CHAPTER 10

In the Beginning

MAN AND NATURE both have left records of climatic change over the past 4,000 years. We have used these records to describe what a changing climate did to people of the Indus valley, Mycenae, Mill Creek, Greenland, and the Sahel—examples especially pertinent to us because they all, even the oldest, are relatively close to our own time. But previous earth history, almost five billion years of it, is worth considering, too.

Numbers like that are abstract to most of us. A useful perspective is to think of condensing the whole of geological time, the whole five billion years, to one year. In such a chronology, with the beginning of the world on January 1, the dinosaurs did not appear until mid-December. Man came along sometime on December 31. And the Indus civilization of 4,000 years ago blossomed near to midnight on that day, in the last minute of the year. The climatic changes that precede the arrival of man should concern us because, as with historical times, the past is a guide to the future. Our climates could again be those of the little ice age, but they could also be those of 10,000 or 100,000 years ago. What were those climates, what caused them, and how suddenly did they appear?

At first, the earth did not even have an atmosphere, and we have no hints of climate before about one billion (1,000 million) years ago. There is some evidence of one or more glacial ages 600 to 800 million years ago. Another ice age about 300 million years ago lasted perhaps 50 million years (National Research Council, 1975).

The very oldest records are not as useful to us as more recent ones; they show less detail, and until 30 million years ago or so the oceans and continents were arranged differently. Once Africa was near the South Pole, and mostly ice-covered. Palms grew in lands that are now arctic. The ocean currents and the land-sea winds were different.

If we move up to one million years ago, practically yesterday in the perspective of five billion years, we can see more clearly what climates are possible, with continents about where they are now. A million years, after all, is quite a large slice of time to work with.

Clues: a million years' worth

Historical records and tree rings are good for a few thousand years at best. Pollen records can cover more time, but have limitations–when glaciers come, no plants grow. And radiocarbon isn't useful for setting dates beyond about 40,000 years ago.

But a number of new techniques, some of them in use only since 1970, are telling us about older climates, especially those of the past million years (National Research Council, 1975). The techniques have some limitations: they can show only broad features and, as with any past climate indicator, it is difficult to calibrate them, to know just what they mean in terms of a thermometer reading or a rain gauge measurement. But several very different indicators agree generally about the past, so that we can have confidence in the overall picture.

The rise and fall of ocean shorelines is one indication of climatic change. To make glaciers that cover large parts of continents, and sometimes are two miles deep, a lot of water has to come out of the oceans. Ice ages are times of low sea level. At the present time the ice caps and other glaciers contain enough water to raise the sea level by about 200 feet. During the height of the last ice age, the sea level was about 400 feet lower yet. In places where evidence of ancient shorelines remains, we can determine the sea level at certain times in the past and can tell approximately how big the glaciers were then.

Ancient soils give another clue to climates of the past million years. A sequence of soil deposits in Czechoslovakia going back almost a million years shows the range of climatic conditions over that time. The soil types include a recurring one typical of cold times, containing the shells of small, cold-dwelling snails. At some other levels are soils of warm climates.

As with all the climate indicators we have talked about, these soil studies must be made in well-chosen places if they are to give useful information. The Czechoslovakian ones came from an area not covered by glaciers, but close to their range. Soil layers deposited over the centuries, in warm times and cold, were not disturbed by glaciers.

The kinds of oxygen found in present-day glaciers and in ocean-floor sediments also tell about climatic changes. Most oxygen has an atomic weight of 16, but some atoms are heavier, with a weight of 18. The proportions of these two in the world as a whole appear to be quite constant over time, and modern instruments can distinguish between them. Since O-16 is lighter, water molecules containing it evaporate from the ocean more readily than molecules with O-18. Clouds, rain, and snow, then, have fewer O-18 atoms than does the ocean water left behind.

In an ice age, a relatively high proportion of the ocean water that ends up in the ice caps by way of snowfall is O-16 water. Correspondingly, the oceans are richer in O-18 than at other times, and this shows up in the fossil shells of tiny marine animals. A time which produced shells with much O-18 was a glacial time, and can be identified as such in ocean-sediment cores, which, in a few cases, go back a million years and more.

Oxygen 18 analysis of ice cores from Greenland and Antarctica gives a picture of the past 10,000 to 100,000 years, more in some cases. Long sequences of climate have been constructed in this way.

In the case of ice cores, the goal is to learn what the air temperature was when the snow fell, not what the ocean temperature was. Whatever the proportions of O-18 and O-16 that go into a cloud, the air temperature affects the proportions that come out as snow. The heavier O-18 molecules presumably condense out at relatively warm temperatures. On a colder day the proportion of O-16 should be greater. In practice, however, the reconstruction of temperatures by this method involves some serious problems. For example, the signature of air temperature could be "forged" by a change in wind. Whatever percentage of O-18 went into clouds, if the clouds came from a great distance most of the heavy O-18 would have fallen already. Clouds from nearby would have relatively more O-18, and therefore would produce snow with more. So a change in where the clouds came from could mimic a temperature change, when none took place. Also, because glacial ice shifts and flows, it is difficult to be certain of getting a smooth present-to-past sequence in a core. At some time, old ice may have overflowed younger ice, mixing up the sequence that scientists assume is there.

Despite such problems, this ice-core technique gives general agreement with several other techniques on both the nature and the timing of major climatic periods. Knowing the date of each piece of evidence is a problem in itself, whatever technique is applied to decipher it. Where a regular pattern is at work, as in tree rings, the scientist can count his way back. Ice cores and ocean sediments are sometimes built like tree rings, with greater deposits during certain parts of the year. But there are other records that develop so uniformly, or so slowly, that it is not possible to count off the years. Ocean sediments may build up an inch or two every millennium. Because of mixing and uneven settling, the thinnest slice of such a core contains material from decades or centuries.

In some such cases, radioactive elements with longer half-lives than radiocarbon can date material older than 40,000 years. In other cases the presence of fossils whose age is known helps in dating.

Sketches of the past

The graphs that follow each give a glimpse of how climates have varied over the past million years or some portion of that time. The million-year record is quite generalized, the 100,000-year record is more detailed, and we have successively more details down to the past 100 years.

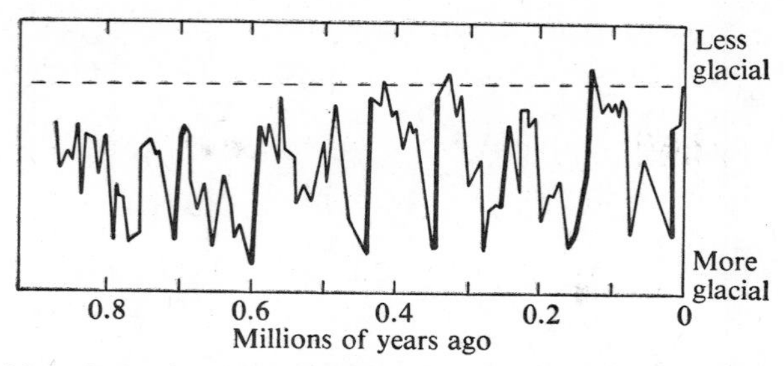

Figure 10.1. Global ice volumes of the past million years, based on O-18/O-16 ratios in deep-sea cores. Heavy lines show times of warming from each of the seven ice ages known within the past 700,000 years. The dashed horizontal line shows that times with as little ice as at present have been rare and brief. Adapted from National Research Council, 1975, p. 130.

Figure 10.1, the record of climate over the past million years, is based on O-18/O-16 readings in the fossils of tiny marine animals deposited on the floor of the Pacific. Throughout this time, ice has covered the earth's poles. The whole time has been a cold one, compared to some earlier ones. Within the million years there have been several outbreaks of ice that we commonly call ice ages.

Many of us picture ice ages as worlds apart from the climates of our lifetimes. That is not quite true. Today, permanent ice covers about 3 percent of the earth's surface. Most of this area is at the South Pole, with Greenland counting for the majority of land ice cover in the Northern Hemisphere. Looking back over the greatest ice ages of the last million years, the total ice cover was not very much greater: at the maximum, permanent ice covered about 9 percent of the earth's surface—a mere 6 percent difference.

Such ice ages have come many times; there have been seven in the past 700,000 years. They have alternated with warmer periods, and each cycle has lasted roughly 100,000 years. To use the term "cycle" is misleading, for it implies regular, dependable changes between colder and warmer times, and in fact, no two of the cycles are identical. But each does show a gradual transition from warmer times to an intense glacial peak, then a relatively abrupt return to warmth.

How abrupt can these changes be—in either direction? No one knows for sure, but some of the records reveal great changes within centuries or less. Remember, the indicators of climate respond to change slowly, some more slowly than others. For instance, a

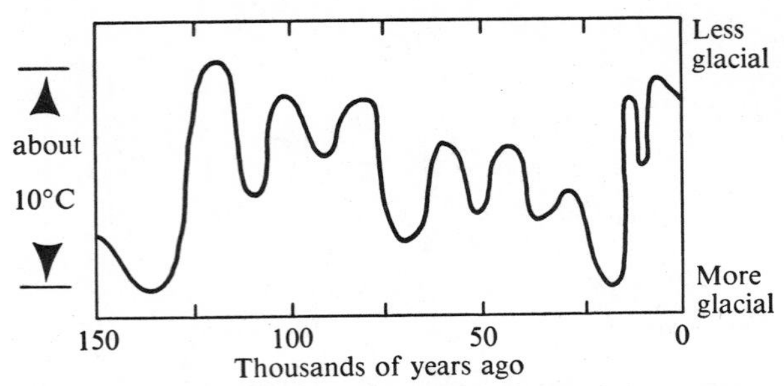

Figure 10.2. Northern Hemisphere temperatures of the past 150,000 years, a schematic rendering on the basis of data from three kinds of sources: marine deposits, pollen records, and shoreline changes. Adapted from National Research Council, 1975, p. 130.

moderate drought that began abruptly would produce a gradual change in the amount of tree pollen: a reduction the first year, even less pollen in the next years as the trees used up reserves, and less still over decades as the forest thinned.

Climatic change comes more abruptly than our records of the past might suggest. We believe some of the shifts from glacial climates to interglacial climates came within a century or so (Bryson, 1974a, p. 759). Right now, we are near the warm peak of a cycle. Over 90 percent of the past million years has been colder than our last several thousand years.

The million-year record gives a perspective, but shows little detail of times close to our own. Figure 10.2 shows that to find a time as warm as the last few years, we have to go back through a long glacial period to 125,000 years ago, during the so-called "Eemian interglacial." In between were fluctuations, but from our perspective the climate ranged from cold to very cold.

When we shift scale again, and this time show the past 25,000 years in more detail, the rises in temperature after the last ice age (figure 10.3) become clear. A number of variations within the past 10,000 years also show.

We constructed graphs (figure 10.4) that begin near the end of the last ice age, about 13,000 years ago. They are based on pollen deposited in a marsh in Minnesota and in a lake in Canada, and show good agreement with the records from other sources (Webb and

Bryson, 1972). The sudden dip in temperature 6,600 years ago matches the eruption of Mount Mazama (now Crater Lake, Oregon), which spread a vast sheet of volcanic ash eastward across the United States and undoubtedly reduced sunlight over a much larger area.

Graphs of the past 1,000 years and past 100 years show some

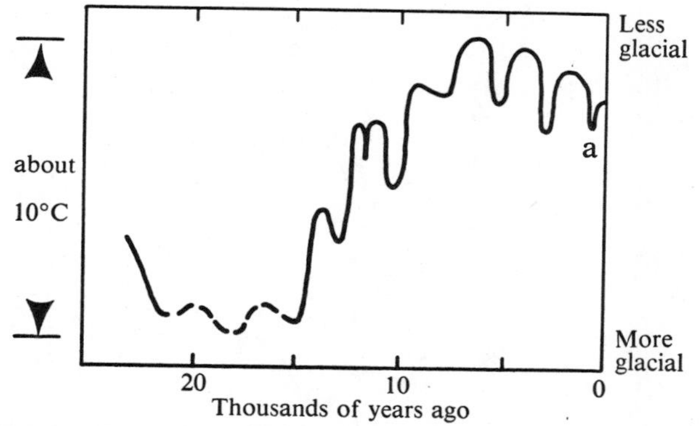

Figure 10.3. Northern Hemisphere temperatures of the past 25,000 years, a schematic rendering on the basis of data from three kinds of sources: pollen records, and tree-line and glacial records. Little ice age is indicated by *a*. Adapted from National Research Council, 1975, p. 130.

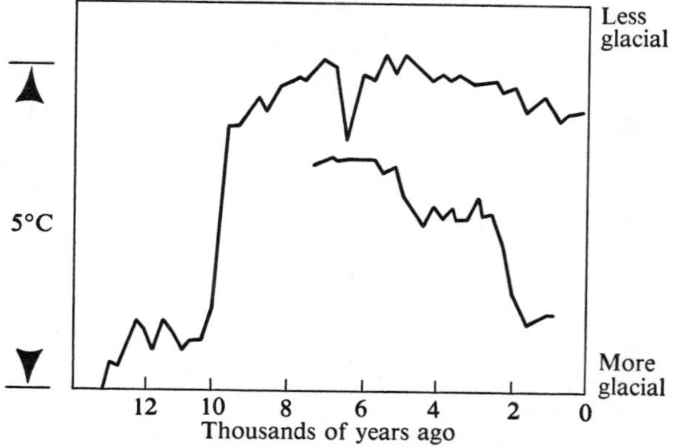

Figure 10.4. July temperatures for the past 13,000 years at Kirchner Marsh, Minnesota (upper curve) and at a lake in southern Manitoba, calculated from pollen data. After Webb and Bryson, 1972, p. 115.

features already familiar. The one for 1,000 years (figure 10.5) is that of Lamb (1966), based on many indications of winter severity in eastern Europe. Note the similarity to the Iceland reconstruction of Bergthórsson (figure 4.2). The change that brought drought to Mill Creek was less significant in eastern Europe than in the Great Plains or Iceland, but it shows.

In the past 100 years, the time of weather instruments, the warmer period that followed the little ice age and lasted until about 1950 (figure 10.6) is plainly visible.

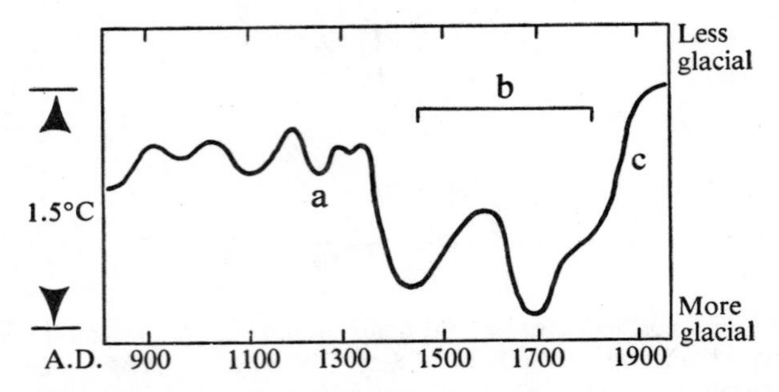

Figure 10.5. Northern Hemisphere temperatures of the past 1,000 years: *a.* Mill Creek cooling; *b.* little ice age; *c.* early twentieth-century warming (since about 1950 cooling has occurred). Adapted from National Research Council, p. 130; based on Lamb, 1969.

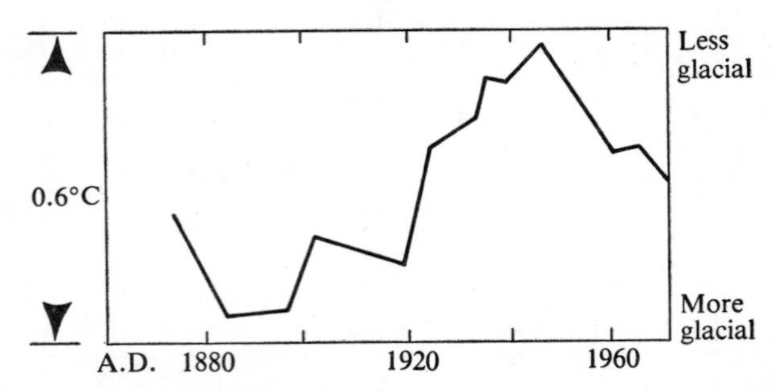

Figure 10.6. Northern Hemisphere temperatures of the past 100 years. Adapted from National Research Council, 1975, p. 130; based on Mitchell, 1963.

The abnormal present

The climate we think of as normal, the 30-year period that weather agencies define as "normal," looks quite abnormal in the perspective of the last 1,000 years. By comparison with longer periods, back to a million years ago, it looks very abnormal.

Toward a consideration of the directions our climate might take, the past makes some interesting suggestions. The conditions which produced these past climatic changes can occur again in the same or other combinations. What has happened can happen.

In recent years our abnormally warm climates started to cool. The Northern Hemisphere has cooled by almost 0.6°C (1°F) since the 1940s. Many past times of cooling have continued much beyond this, but others have instead shifted back to greater warmth after a time. The pattern of past climate is not regular enough to reveal what is coming next.

Are we then left with only a knowledge of the climates that are possible, and with no hint of which ones will turn up next? Not quite, for climate is not random. We can consider some possible causes of climatic change, and see where those forces or conditions might be leading us now.

As we look for physical causes, though, it is worthwhile to keep in mind the lengths of time between past ice ages–to remember past patterns, irregular as they have been. As we said, about 90 percent of the past million years has been colder than our times. Warm interglacial periods like ours have blossomed about once in every 100,000 years over the past half-million years. And these warm interglacials have each lasted between 8,000 and 12,000 years.

The present one has already lasted about 10,800 years (Bryson, 1975a).

CHAPTER **11**

How Climate Changes

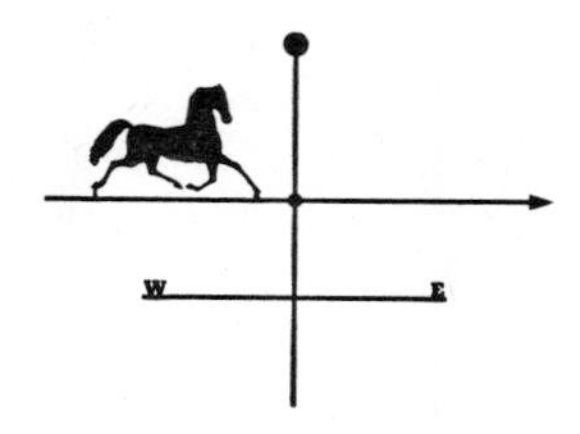

ALL THOSE CLIMATIC changes we described in the last chapter, and in other parts of this book, have some cause—or causes. The formation of continental glaciers, thousands of feet thick and thousands of miles wide and long, is a response to some physical condition. So is a less dramatic expansion of the westerlies, like the one that brought drought to Mill Creek, and so is the warming of this century followed by the present cooling.

The causes are being sorted out slowly, as is the record of changes.

Before looking at how climates change, let's consider the cause of a major feature of the world's climates: the westerlies.

Sunshine: how much and where

The main driving force for the earth's atmosphere is the sun, which heats the earth unevenly and therefore makes the air move. Because of the earth's spherical shape and tilt on its axis, the tropical regions receive, in the course of a year, about two and a half times the sunlight per unit area that the polar regions get (Lamb, 1966, p. 22). That is the single most important fact in determining our climates. The tropical regions are warmer than the polar regions. Besides this north-south difference, there exist east-west temperature differences around the earth at any given latitude. These are due to the different ways land and sea heat up when the sun shines on them.

These differences in temperature of large sections of the earth (both north-south and east-west differences) produce large convection currents. And in particular, the north-south difference drives the westerlies. But there is another factor, the earth's rotation, which is also partly responsible for them.

Why the westerlies flow

The earth spins once every 24 hours. Since the earth is about 25,000 miles round at the equator, a point on the surface there is moving eastward at a little over 1,000 miles per hour. At Chicago, the eastward speed is about half as great, and of course the North Pole doesn't move eastward at all.

A mass of air which is moving neither east nor west *relative to the surface* has a true eastward velocity the same as the surface, and a corresponding eastward momentum. If the mass of air moves to a different latitude, it will tend to retain its original eastward momentum, though losing some of it, primarily by friction with the surface.

The uneven heating of the earth, as we have mentioned earlier, causes air to rise in the tropics and flow northward (and southward) at high altitudes. As it moves away from the equator, its excess eastward momentum *relative to the earth's surface* makes it appear as a west wind to an observer beneath it.

This, in a very simplified version, is why there are west winds, and weather moving west to east, in the mid-latitudes. Air flowing into the mid-latitudes from the tropics has an eastward momentum, producing the westerlies.

The world in a dishpan

We can get to more of the essentials by leaving the complicated spherical world for a moment and describing a simple model, constructed by the meteorologist Dave Fultz (1959, 1961). Fultz used a pan of water as a model of the Northern Hemisphere. He considered the North Pole to be at the center and the equator to be the outside rim. Looking straight down on the pan, then, he had the view of someone suspended far above the North Pole.

He then applied heat to part of the pan to simulate the uneven heating of the earth. This produced a circulation equivalent to air rising in the tropics (the outside of the pan), flowing toward the center, and then down.

Then Fultz added rotation, because the earth rotates. He put the pan on a turntable and rotated it as he applied heat. A spaghetti-pot circulation of the kind we earlier described did show up for conditions simulating the tropics. But for conditions simulating higher latitudes a meandering circuit of westerlies developed, a looping flow around the dishpan's center much like the westerlies in the atmosphere. Fultz had shown how and why westerlies flow in the atmosphere.

The earth is not a dishpan. But every model is an approximation of reality, and this one turned out to be an excellent analogy. It behaves like the atmosphere in important ways; this is confirmed as more is learned about actual patterns of climate through weather satellites and other modern instrumentation.

The Fultz experiment not only showed that heat plus rotation makes westerlies, it revealed crucial facts about how climates change. It showed that as the equator-to-pole temperature difference is increased, the westerlies expand. It is the difference, the temperature gradient, that is important. The dishpan experiment shows that if you keep the equator at one temperature while cooling the pole, the "westerlies" will expand.

This happens in the atmosphere too. Polar regions go from continuous sunlight in midsummer to continuous darkness in midwinter, cooling greatly. The tropics don't undergo much change in temperature, and the westerlies expand.

Thus, anything that changes the net amount of sun energy received by the earth will change climates, and anything that changes the *distribution* of sun energy on the earth will also change the westerlies and consequently change climates.

And temperatures from bottom to top

There is another factor that determines the westerlies' location and pattern, and therefore the weather in North America, Europe, and the monsoon lands. It is predicted by meteorological theory, and its effects can be observed in the atmosphere under certain conditions.

This factor is the vertical temperature difference, the difference between the temperature at ground level and that higher up. Particularly important is the difference that exists in the temperate regions—the middle latitudes.

As a general rule, temperatures drop as you move upward a few miles into the atmosphere. The difference that you find, however, varies from time to time and place to place. The connection between the vertical temperature difference and the westerlies is not easy to visualize, but it has been determined that an increased temperature difference makes for expanded westerlies (Flohn, 1965). In other words, if the air near the ground is heated or the air at high altitudes cooled, the westerlies tend to expand.

We can illustrate this by looking at the westerlies in two months, one in spring and one in fall, with identical equator-to-pole temperature differences. We might expect that, if the westerlies depended only on that temperature difference, they would have the same pattern in both months.

But they do not. In the spring month the westerlies are observed to be larger, more expanded, than in the corresponding month in the fall: in spring the sun has begun to warm the ground and lower air, but the air high above is still very cold from winter, resulting in a big temperature difference between the surface air and the upper air. In the fall month, with an identical equator-to-pole temperature difference, the westerlies are less expanded. As the entire atmosphere begins to cool in the fall, the surface air cools faster than the upper air at first, making the temperature difference between them smaller (Bryson, 1973, p. 369).

We have, then, two relationships which determine changes in the westerlies: the equator-to-pole temperature difference and the bottom-to-top temperature difference. If these increase, the westerlies expand.

Moreover, when the westerlies do expand, their configuration changes. They frequently make large digressions to the north and the south. With more looping westerlies, weather patterns move to the east more slowly; a given pattern persists in an area a relatively long time. Looping can bring a warm April, followed by a frosty May, because warm loops from the south flow over a region, making it warm, and a subsequent shift in the pattern can bring a cold flow down from the north over the same place.

Step by step

There is another fact about the westerlies (and the dishpan experiment). In both cases, the westerlies must make a whole number of loops or waves in their circuit around the pole. They make three loops, or four, or five—but not three and one-half or four and two-thirds. The flow of westerlies must "catch its own tail." The flow is affected somewhat by the earth's high mountain ranges, but whatever the exact path, there must be a whole number of loops.

The crucial implication is this: when the number of loops in the westerlies changes, it must change from one whole number to another. The conditions that bring on a change (north-south and top-bottom temperature differences) may change gradually. But as they do, there comes a point at which the westerlies change abruptly, from one pattern to another. The dishpan experiment showed that after establishing a certain wave pattern, say three loops, it is possible to change the heat gradually and the three loops will remain. But change it enough, and the three will flip to four.

This *suggests* why climate can change quite suddenly, as we have said throughout this book. An ice-age climate can take hold in a century or less; a Mill Creek drought can develop in a decade or so; monsoon rains can disappear from an area within a few years. The relation of these sudden changes to the "flip" in westerly patterns can be seen in an example from the monsoon lands.

We said earlier that the monsoons come to India when the westerlies switch from south of the Himalayas to north of those mountains, allowing the summer monsoon from the Arabian Sea to penetrate into the continent. And that switch is associated with one of the steps in the contraction of the westerlies as the hemisphere warms in the spring.

It's possible that the westerlies may not make that particular change in a given year, or may make it too late for the rain to help the Indian crops. Perhaps one year the earth's temperature patterns are just favorable enough for the change in the westerlies that brings the monsoon to India. Perhaps the next year the temperature patterns are different–a seemingly insignificant difference that could have drastic results indeed. And as we discussed in chapter 7, there is a correlation between expansion of regions of arctic air and Indian monsoon failure.

So climate patterns change stepwise, not in a smooth, gradual progression. A small change in the north-south or top-bottom temperature differences can bring a new pattern of the westerlies for a part of the year, and hence bring great changes to the activities of people.

Causes of change

In the brief survey that follows we will describe some mechanisms and conditions that have been proposed as causes of climatic change. They fall into two general categories: (1) internal oscillations within the climate system, and (2) external influences–such as continental drift; changes in the output of sun energy; changes in the reflectivity of the earth's surface, which change the amount of sun energy reflected back to space; changes in the earth's orbit, which in turn change how much sun energy the earth receives or which influence the atmosphere directly; and changes in the composition of the atmosphere that affect incoming sun energy or create a "greenhouse effect."

The first category, internal oscillations, or changes within the earth's climate system, involves processes within the air, water, ice, and land components that are affected by climatic change and may in turn themselves cause climatic change (National Research Council, 1975, pp. 23-25). Such internal processes have not been proved to cause climatic change. But here is one projected example.

Suppose a time of warm climate exists, for whatever reason. In such a time, a good deal of water would evaporate from the oceans, making many clouds. Clouds reflect sunlight away, and can thereby bring on a cooler time, one in which ocean evaporation would be less. There would be fewer clouds, and temperatures might increase again for a time.

If such oscillations do exist apart from external causes, however, they do not resemble the swing of a pendulum. The patterns are irregular; the complex interactions of the atmosphere cannot be reduced to clockwork. Some rhythms do show up in climatic data, but they are not dependable. For instance, some climatic variations have a tendency to repeat about every 2.2 years (or about every 26 months). Climatic patterns of any given month are more likely to resemble those of 2.2 years before than chance alone would suggest. Or sometimes a seven-day weather cycle appears—you may have noticed times when it rained every weekend. Climatologists have noticed those too, but the duration of such cycles, like that of most weather cycles, is not dependable.

Two points about such rhythms should be kept in mind. First, as we have said, they are not a dependable guide for the future. Second, to the extent that they do exist, they might not be the result of internal causes, but rather might be a response to some cause outside the weather system itself.

The external mechanisms which constitute the second category—mechanisms not themselves affected by climatic change—clearly can change climate. The changes brought about by such mechanisms are not continuing cycles of action and reaction. In the pages that follow we consider a number of the external influences that might be causes of climate change.

Continental drift

One external cause of climatic change is changing earth geography. We now know that the continents move in relation to each other—this is called continental drift. Not until 30 million years or so ago did continental movement permit something like the arrangement of ocean currents we know today.

Continental drift can affect the climate of a continent in two ways. First, continents move from one latitude of the earth to another, and hence experience different climates. There is evidence of glacial cover over central Africa, at a time when Africa was near the South Pole. Second, recall how the distribution of land and sea determines where the monsoons, and other continental winds, can blow. Also, ocean currents have their boundaries firmly fixed by land masses. The Gulf Stream flows toward Europe, making it relatively warm, partly because of the way the Atlantic Ocean and its borders are shaped.

The continents are drifting today, but only inches per year or less, too slowly to bring climatic change that should be of concern to us, or even to our descendants a few thousand years from now.

A changing sun?

Another cause of climatic change could be a change in the sun's output over time. But we have no proof that the sun's output changes. To test the idea, we need long-term measurements of the sun's intensity, precise to a fraction of 1 percent. But we don't have measurements so precise, so the question remains open (Schneider and Mass, 1975).

What about the possible role of sunspots in climates? Scientists and others have claimed a connection exists—have claimed that sunspots may, in fact, be directly related to the sun's output. Some of the early studies of tree rings were efforts to find cycles of about 11 years, the average time between peaks of sunspot activity. But cause and effect have not been established. One problem for the sunspot advocates is that an 11-year earth temperature cycle, if it does exist, is very difficult to identify.

Suppose, for instance, that a rise of sunspot activity really is tied to a slight increase in sun energy coming to us. And suppose incoming sun energy jumps to a higher level. It will take a decade or so for that increased energy to heat the vast oceans to something approaching a new equilibrium temperature (Bryson and Dittberner, 1976). But by that time, the sunspot activity is falling again. So any sunspot signatures that do exist in earth climates are obscured.

Some scientists are investigating the possible connection between sunspots and climatic periods much longer than an 11-year cycle (Schneider and Mass, 1975).

Reflectivity

A change in the amount of sunlight reflected back could also change climate. The reflectivity of the earth changes when land masses become lighter or darker from weathering, when vegetation cover grows or shrinks, and when snow and ice cover changes.

Such changes in reflectivity are important for climates, because the *net* energy balance of the earth–energy coming in less that going out–determines our planet's temperature. Reflectivity varies greatly. The dark seas absorb several times more energy from the sun than do clouds or snow-covered fields.

Our orbit

Yet another way to change the net sun energy received by earth would be to change the earth's orbit.

The average distance from sun to earth varies somewhat. So does the shape of the orbit; it is more elliptical at some times than others. This means it is closer to the sun in certain seasons, so the distribution of sunlight is changed. And the earth wobbles on its axis. That also changes the distribution of sunlight.

Several of these earth orbit variations take place over many thousands of years. One complete wobble in the earth's tilt, for instance, takes about 40,000 years.

A Yugoslav mathematician, M. Milankovitch (1920, 1930, 1938), calculated the changes in incoming radiation, latitude by latitude, that result from these variations, and developed a model that related the orbit variations to past ice ages–a monumental work for the days before electronic computers. Milankovitch was not able to take into account some important factors like the role of the atmosphere in moving heat about, so he oversimplified the situation somewhat, and his ideas were disregarded for many years.

But climatologists are coming to realize that he found something important. Recent work, including some at the University of Wisconsin-Madison, indicates that earth orbit variations might indeed cause significant changes in climates. We investigated the changes in radiation and the changes in the westerlies that could result, and concluded that such changes, persisting over thousands of years, might well produce ice ages (Broecker, 1968; Kutzbach, Bryson, and Shen, 1968).

Since the long-term variations occur regularly, they can be calculated for the past and the future. These calculations match up well with past ice ages, especially with indications of past ice ages as revealed in sea bottom records. It is possible to predict the future sequence of these orbit changes, because they are regular. The future orbit patterns suggest that, if orbit variations were the only cause of climatic change, a new glacial period would begin within a few thousand years.

Earth orbit variations, then, may explain climatic change that takes place over tens of thousands of years. They may be important in bringing the succession of glacial and interglacial periods the earth has seen in the past million years or longer.

Still another earth orbit variation may explain shorter-term climatic variations. While work on this subject is still in the early stages, it is possible that a small wobble in the earth's rotation–one that repeats on the scale of years rather than millennia–may directly produce certain patterns in the atmosphere that can be predicted (Bryson and Starr, 1976).

What the air contains

Now we come to a consideration of changes within the earth's atmosphere. Substances in our atmosphere can reflect the sunlight away, or hold in heat like a blanket.

The composition of the atmosphere can change for a number of reasons. We will spend the next chapter on two agents with great potential for changing climates rapidly and significantly by changing the atmosphere: volcanoes and humans.

CHAPTER 12

Pollutants in the Air

ANYTHING IN THE atmosphere which can change the net amount of sun energy the earth receives, or which can change where sun energy is concentrated, will affect the westerlies and will therefore change climate.

Dust and other substances can alter the atmosphere rapidly: within a human generation or less they can change climates significantly (Bryson and Dittberner, 1976). In a world where a change in our climate can bring mass starvation, we need to consider the ways pollution can change climate. Air pollutants can change the westerlies by changing the equator-to-pole temperature difference, or by changing the bottom-to-top temperature difference. An increased temperature difference in either direction, you will recall, expands the westerlies.

We'll consider one substance that can change the equator-pole difference and one that can change the vertical difference. First, bottom-to-top.

People who live in a greenhouse

Within the past decade, the "greenhouse effect" of carbon dioxide in the atmosphere has become famous. In a greenhouse, sun energy comes in through the windows. When it strikes the surfaces inside, it warms them. The wavelength of the radiation is changed—"light," you might say, is changed to "heat." The greenhouse windows pass light more readily than heat. Some of the energy that enters is trapped, and the greenhouse gets warm.

Certain substances in the atmosphere do the same thing, letting sunlight in but preventing some heat from radiating back out to space. Water vapor is one such substance, and this is why humid nights cool off more slowly than very dry ones.

Carbon dioxide (CO_2) has a similar effect (Manabe and Wetherald, 1967). Some CO_2 comes from natural processes, including breathing. It is not poisonous like carbon monoxide; green plants need it to live. But CO_2 in the atmosphere has increased about 10 percent in the last century, and most of this increase seems to be due to fuels we have been burning in our age of industry and the automobile.

Carbon dioxide, by itself, like a greenhouse window, should warm up the earth. Some people have feared a global heat wave and the melting of ice caps. But before jumping to that conclusion, we need to consider just what carbon dioxide can do in the atmosphere. The temperature of the whole system, earth and atmosphere, depends on the balance between the energy coming from the sun and that energy *reflected* back away. Carbon dioxide does not change the amount of energy coming from the sun, and it does not reflect the sun's energy back to space. Therefore it can't change the average temperature of the whole system—can't make it warmer. It does, however, warm the lower parts of the atmosphere; so the upper atmosphere must get colder (Bryson, 1973).

This, in combination with our earlier argument about an increased bottom-to-top temperature difference in the atmosphere, has an important implication. We discussed how an increase in this temperature difference could expand the westerlies. The implication, then, is that increases in carbon dioxide can work to expand the westerlies (Bryson, 1973, p. 370; Bryson, 1974a, p. 756).

A veil of dust

Carbon dioxide does not change, but redistributes, the total energy received by the atmosphere and earth. If we could turn the sun's intensity up or down, *that* of course would change the total energy—and so would a change in how much sunlight is reflected back.

Some things do change reflectivity. Photographs from space show that clouds are reflecting sunlight back—they are white. If something changed average cloudiness significantly, the earth's overall temperature would change. But there is some feedback, some self-regulation, in the process of cloud formation, as explained earlier. Greater cloudiness cools the earth somewhat, less water evaporates from oceans and land, and cloudiness diminishes. Recent research indicates that this self-regulation prevents changes in average cloudiness from seriously affecting overall temperature (Cess, 1976).

There is another substance in the atmosphere that reflects back sunlight, and that does not regulate itself: dust.

Dust can cool the earth by reflecting away a significant portion of the sun energy that would otherwise be received. Dust from volcanoes has brought times of cold climate. In a volcanic eruption, molten rock finds its way up through the earth's surface. It may flow out, or may be blown violently out. The violent explosions can throw enormous amounts of debris high into the atmosphere; these eruptions are the ones that influence climate.

The chief evidence that volcanic dust affects climate comes from a few incidents in the past two centuries and from the development of careful comparisons between volcanic eruptions and climates. H. H. Lamb (1970) has done some important work on this subject.

In 1815, the Tambora eruption in Indonesia threw a veil of dust into the stratosphere that made 1816 famous as "the year without a summer." Throughout the middle latitudes of the Northern Hemisphere, temperatures averaged about 1°C (1.8°F) below normal. In some parts of England the summer was perhaps 3°C (5°F) colder than normal, and certain regions had rain on all but three or four days from May to October. New England and eastern Canada got widespread snow in early June, and frosts each month of 1816. As in other years of the little ice age, crops were killed by frost, or failed to ripen and rotted in the field. Times were especially bad in northern Ireland, and Wales had food riots.

The fact that one volcano half way round the world could have such consequences on both sides of the Atlantic is another indication of how sensitive our climate is. The events of 1816 show how serious a one- or two-degree change in hemispheric temperature can be.

Just a few years earlier, from 1811 to 1813, other volcanoes had produced dust that drifted round the hemisphere, too. After marching into Russia in the summer of 1812, Napoleon retreated several months later–in the face of the Russians and a bitterly cold winter. It has sometimes been speculated that the winter of 1812 was in part a product of the volcanoes.

By the nineteenth century, world travel was extensive enough and fast enough so that the connection between an explosion in Asia and the hazy red sun that appeared in North America and Europe a few months later could readily be made. The connection between the dull sun and the winter-like summer was even more obvious.

Lamb and others have looked in detail at other volcanoes and climatic changes. Some single explosions, like that of Tambora, have cooled climates; but waves of volcanic activity also have appeared over centuries. The three most recent ones, according to some authors, were about 3500 to 3000 B.C., about 500 to 200 B.C., about A.D. 1500 to 1900 (Lamb, 1970, p. 482).

What sets off volcanoes?

It has been argued that these times of high volcanic activity were also times of cool climate, and that they support the theory that volcanic dust cools the earth. The last of the three periods of volcanic activity corresponds very well with the little ice age.

But if we can make the case that episodes of great volcanic activity bring cool climates, another question comes up. Why do certain times have much volcanic activity? Does something from time to time trigger eruptions around the world? There are some possible explanations.

Perhaps movement of continents causes a buildup of stresses over large areas, which can be released in volcanic–and earthquake–activity at certain times. Other stresses may be important. As ice ages come and go, a great weight of water is moved between oceans and glaciers that cover the land. Lamb suggests (1970, pp. 482,

495) that these shifts may set off volcanoes. The theory is that after an ice age ends and the land is relieved of its load, events are set in motion that several thousand years later will trigger volcanoes, blocking out the sun and bringing another ice age.

What volcanoes do

Whatever the cause of volcanic activity, volcanoes do influence climate. The greatest eruption of modern times, the Krakatoa eruption of 1883 in Indonesia, is another clear case of cause and climatic effect.

Krakatoa threw dust into the stratosphere (which begins about five or ten miles up)–dust that circled the globe. In August, in the main explosion, perhaps as much as 13 cubic miles of material was blown out (Wexler, 1952, p. 74), and enough of it got carried such great distances that red sunrises and sunsets were seen far away and for months afterward. Periodicals of the day, including *Scientific American* (1883), contained reports of unusual sky colors: a green sun in India, a rainbow in a clear sky and brilliant sunsets in the United States, and peculiar sunsets in Peru. One correspondent attributed the sunsets to a "swarm of meteors." After several weeks, a reader made a connection between these events and the previously reported Krakatoa explosion. *The New York Times* reported on November 28 that the fire department in Poughkeepsie, New York, had been called out the day before because the brilliant sunset was mistaken for the glow of a fire in the western part of the city.

Those colored skies were one obvious sign that sunlight was being scattered by the volcanic dust. But how does such a hemisphere-wide or worldwide spread of dust from an eruption affect the total amount of energy flowing from the sun and from the earth?

1. Lamb argues (1970, p. 461) that dust has a "reverse greenhouse" effect. Carbon dioxide, we pointed out earlier, lets in most of the sun's "light" but blocks "heat" that would otherwise escape from the earth. Dust does the opposite, Lamb says. It blocks out sunlight, but has less ability to keep the earth's heat in. This reverse greenhouse effect depends on the size of the dust particles. Large particles can *hold in* some heat; for example, the dust over the Indus valley contains some larger particles and thus keeps nighttimes warm. Particles small enough to be carried far up into the

atmosphere, and to stay there for months or longer, have the greatest reverse greenhouse effect. So dust works to cool the lower atmosphere, rather than warm it as carbon dioxide does. But since dust, unlike carbon dioxide, actually changes the amount of energy the earth-atmosphere receives, it also lowers the overall temperature of the system. It does not merely redistribute the energy.

2. Dust in the atmosphere tends to cool the high latitudes more than it does the tropical regions, no matter where it enters the atmosphere. Some dust is carried poleward, over a period of weeks and months, by the high-altitude flow of air from the tropics. More important, even if dust were distributed evenly throughout the atmosphere, the poleward regions would be more shaded by it than the tropics. Sunlight takes a nearly vertical path through the atmosphere in the tropics, but away from them comes in at an angle, and therefore has a longer path through the atmosphere. If the atmosphere is dusty, the sunlight has a longer path through the dust–and is diminished all along the way.

Lamb has designed a dust veil index for calculating the relative importance of various volcanic eruptions on climate (1970). This and other work relates climate to volcanic dust, both theoretically and historically. But as Lamb points out, volcanoes do not explain all climatic change. For instance, the coldest times in the little ice age do not seem to match up with the greatest concentrations of dust. Severe years or seasons have appeared whether or not a lot of volcanic dust was present. The most recent cooling, which began around 1950, seemed to be one of those times that cannot be explained entirely by volcanoes, because everyone thought volcanoes had been very quiet. But recent work at the University of Wisconsin-Madison (Hirschboeck, 1976) has uncovered more records of volcanoes, and the last 25 years have not been as free of eruptions as we had thought. Still, the present cooling had already begun when the new period of active volcanoes started, and another force must have been at work.

How people change climate

Dust today also comes from more sources than volcanoes. It comes from industrial smokestacks, from wind over barren soil, from automobile and airplane exhausts, from farm tractors, and from the fires set by Third World farmers to clear their land. The last category is more significant than you might think. The smoke of slash-and-burn agriculture accounts for perhaps one-tenth to one-fifth of all the material humans put into the atmosphere (Bryson, 1974a, p. 758; Bach, 1976). In addition, smoke particles are small and stay up longer than some other particles.

People, by means of industry, agriculture, and their other activities, put approximately 500 to 600 million metric tons of material into the air every year (Bach, 1976). If that seems like a very high figure, consider that it amounts to only ounces per person per day. But more important than the amount going in is the dust-loading of the atmosphere: the amount of dust suspended at any one time. One estimate (Bryson, 1974a, p. 758) is that all human activities today produce a dust-loading on the order of 15 million metric tons–about the same as moderate volcanic activity.

All this dust can decrease the average temperature north of the tropics in the Northern Hemisphere, and thereby cause the westerlies to expand.

We are suggesting that people can play an important–though generally unintentional–role in climatic change. Can people really affect such a massive collection of forces as climate?

Consider not the globe or the hemisphere, but one large city. Anyone who has breathed or seen a large city's air knows that humans leave their mark on it. Automobiles, power generation, and industrial activity all add material to the air. Some of our contributions are sulfur dioxide, carbon dioxide, carbon monoxide, hydrocarbons, nitrogen oxides, and particulates. Most of these also come from other, "natural," sources–for example, when dead leaves decay they give off carbon dioxide; wind blew up dust before humans scratched the earth's surface. But people do add large amounts of these materials, as much as nature or more in some cases.

Air pollutants are one characteristic of cities (though the nature and amount of pollutants vary). Some other characteristics that affect the city climate are large paved areas and tall buildings–as opposed to green and open spaces; storm sewers that get rid of precipitation quickly; and high use of energy. What might the total effect be on the climate of the city?

There is an important effect on temperature (Bryson, 1972). A great amount of energy is used in cities, and most of it warms up the city air. Heat from car engines and home furnaces, and heat that is pumped out of buildings to cool them in the summer, doesn't "disappear." It warms the outside air.

Electric generation is only about one-third efficient: of the energy brought to a generating plant in the form of fuel, about one-third goes out through the wires and two-thirds goes up the chimney, into the atmosphere, along with pollutants. The heat from the houses does too, after a slight lag. Even people give off heat–in normal activity, it is perhaps 150 watts apiece. In New York City in mid-winter, the city's heat production is more than twice the heat it receives from the sun (Bryson, 1972, p. 136).

All this heat makes cities a little warmer than the surrounding countryside. But even without these sources, the cities would be warmer. Storm sewers take away water that would otherwise evaporate, directly or through plants, and use up some of the sun's heat. The pavement and buildings and city air get hot, and since concrete, asphalt, and steel store heat more easily than green leaves and loose soil, the city stays warmer at night. Asphalt and other dark substances in cities reflect little sun energy. Therefore cities absorb energy at a high rate.

Although the sun may shine wanly through the turbid city air, the net result of all this is usually a city warmer than the countryside, by 1°-2°C (about 2°-3°F) or more over the year. We have already seen how important a temperature change of a few degrees can be.

This change causes the air over the city to stay in the area and circulate in a dome-shaped pattern–a "dust-dome" (figure 12.1). The warm central city is like a beach or island that's warmer than the nearby water at night. So the air over this "island" rises and a

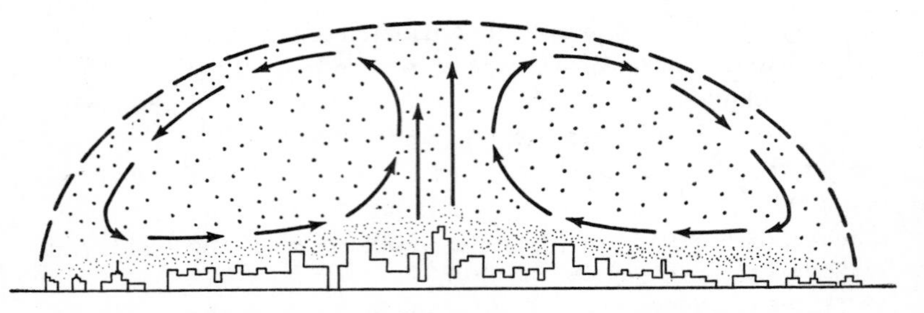

Figure 12.1. A city dust-dome. Arrows indicate direction of airflow.

"sea breeze" flows to it. But the city is likely to be warmer in the day also, so the air continually rises, and flows out to the suburbs and farmland, where it sinks and flows back along the ground to the city—the spaghetti-pot circulation again.

Along a beach, in a city, or throughout a hemisphere temperature patterns affect air movement patterns. In the case of a city, the air contains dust—especially in the central city. As the warm air rises there, some of the dust is heavy enough to fall back rather than go up and out. Some of it circulates around and around the dome. The dust gets thicker and thicker, especially if a layer of cool air over the dome (an inversion) holds it down. A good wind will blow the dust dome away. When the dome does break a plume of dirty air stretches out. On a flight across the North Atlantic, Bryson observed a plume of city air from New York to Iceland.

The dust of cities extends over, and can be seen over, large regions. And there are greater sources of dust; the Indus valley desert, perpetuated by humans, is one. Slash-and-burn agriculture, over southeastern Asia, the Congo, northeastern Brazil, Central America, and other places, puts millions of tons of particles into the atmosphere. It can spread a 15,000-foot-deep layer of smoky air over many hundred thousands of square miles.

Dust in the Northern Hemisphere has increased in our century, especially in the last three or four decades. Since 1940 the lower atmosphere has become about 2 percent less transparent (Bryson and Dittberner, 1976).

Can dust change climate?

There is much disagreement about what the increase in dust can do. The matter is certainly complex. Dust reflects back some sunlight, but it also absorbs some heat, which in turn radiates in all directions. And dust can reradiate some heat back to earth, too, as it does in the Indus valley.

But Lamb's correlation of volcanic activity with cool climate is strong evidence that dust has a net cooling effect. Some people who agree that it does still question whether the cooling is large enough to be important, or whether it might not be offset by the carbon dioxide greenhouse effect anyway.

In the final analysis, what is the result of dust, or dust plus carbon dioxide, in the atmosphere?

An equation for the relationship between dust and carbon dioxide in the atmosphere, and surface temperature in the Northern Hemisphere, has been developed at the University of Wisconsin-Madison (Bryson and Dittberner, 1976). This model points to dust as the explanation for the greatest part of the temperature variation during this century—by far. It indicates that dust accounts for perhaps 90 percent of the temperature variation, while carbon dioxide accounts for only about 3 percent.

Our conclusion is that the net effect of man's burning of fossil fuels, his slash-and-burn agriculture, and his other activities which produce both carbon dioxide and dust, is to reduce temperatures.

More important than the cooling itself is the resulting change in the westerlies to patterns that are more expanded and more looping.

Drought in the monsoon lands and elsewhere, shorter growing seasons in the world's main food-producing areas, and more highly variable weather around the globe: manmade dust, we are convinced, is implicated in all of these.

During the twentieth century we have come to recognize that humans must be added to the list of forces that can change climates around the world.

CHAPTER 13

The Lessons of Climatic History

WE CAN LEARN from Mycenae and Mill Creek, from Iceland and the monsoon lands, because what has happened to climate in the past can happen again. And what has happened?

First and most important, we know that climate is not fixed. On a long time scale, it has varied from glacial, with vast continental glaciers, to nonglacial times like the past 10,000 years. Climatic changes on the scale of centuries or decades have also produced significant ecological changes.

Second, climate tends to change rapidly rather than gradually. Small changes in temperatures can alter the flow of the westerlies, and therefore change climates around the world, from far north regions like Iceland to the tropical monsoon lands.

Since the westerlies must change in whole numbers of loops, they can bring abrupt changes in climatic patterns in response to a small and gradual temperature change in the hemisphere.

The change from a glacial to a nonglacial climate may take less than a century, though plant communities and the environment generally may take a much longer time to complete their response to the new climate. Smaller, but still important, changes of climate may occur in a few decades. When you consider that a full ice age may result from a mean hemispheric cooling of 6°C (10°F) or so, a temperature change of one or two degrees begins to sound significant. In an area like Iceland, with a climate marginal for agriculture, the response to climatic change is very great. A decrease of about 0.6°C (1°F) in Iceland's average growing-season temperature brings down hay yields by 15 to 17 percent (Bryson, 1974c).

Our discussions of the "small" climatic changes that brought glacial advance in Europe, of the sensitivity of wine harvest dates to temperature, and of the small changes in the pattern of the westerlies that can produce a drought in Mycenae or Mill Creek indicate how fragile our climate is.

Third, climatic changes can change civilizations–can affect their survival. Our analysis of the radiocarbon dates reported by other scientists, investigating the past for many different reasons, provided certain dates during the past 10,000 years when widely separated cultures rose or fell rapidly, over a period of decades or less. One such time was about A.D. 1200, when the Mill Creek drought struck. Another was in the eighth century A.D., when the Viking expansion began; still another was 1,200 years or so before Christ, when Mycenae declined.

Synchronous changes at many points on the globe, at these times and at some others, indicate that there must have been global causes. And pollen analysis suggests that in fact climates did change at about the times we have been discussing (Wendland and Bryson, 1974).

Fourth, our present climate, which we think of as normal, is not normal in the longer perspective (Bryson, Ross, Hougas, and Engelbert, 1974). For instance:

> Since 1880 only half of the decades in the Northern Hemisphere have been as warm as or warmer than 1960–70.
>
> Our evidence indicates that since A.D. 1700 all 30-year periods have been colder than the 1931–60 period.
>
> About 90 percent of the past million years has been colder than our own time.

So our present Northern Hemisphere climate is unusually warm. But since about 1950 it has been cooling off irregularly, moving toward temperatures more typical of the climatic history we know. In the Northern Hemisphere, this recent cooling has approached 0.6°C (1°F). It began around 1945 or 1950 and shows up in records from the central and eastern United States, as well as from other parts of the hemisphere. The growing season in England has been shortened roughly two weeks. As the high latitudes have cooled, the area of snow and ice has increased, reflecting more sunlight and thus probably increasing the cooling (Kukla and Kukla, 1974).

During cooler periods, our analysis of past climates indicates, there is greater variability in week-to-week and year-to-year weather. The looping patterns of the westerlies that come with cool climates bring slowly moving weather patterns, and predominantly north and south airflows. In the 1930s the westerlies for about half the days in a given year could be classed as having looping patterns, in the 1950s about 60 percent, and in the 1960s over 70 percent, according to figures derived from work published in 1968 by the Soviet climatologist Dzerdzeevskij. One result of an increase in looping flow can be warm early springs followed by late frosts. Variable fall weather is also likely, as are prolonged wet and dry spells.

Fifth, a cooling and variable climate brings serious problems for world food production. The present cooling is concentrated in the poleward regions, the high latitudes, and when these areas cool, the monsoons tend to fail. The high latitudes have been cooling for nearly three decades, and the hungry half of the world is concentrated in the monsoon lands.

A continuation of the cooling trend, or even a stabilization at today's temperatures, means problems for temperate zone agriculture as well, in the form of slightly shorter growing seasons, less summer warmth on the average, and unreliable rainfall. For crops that have been highly developed to match the climates of recent decades—"normal" climates—any change in conditions brings a lower yield.

There is more to our lives and to the course of nations than climatic change. Many events cannot be explained in terms of climate. But our environments do set some limits. Our climates do have a great influence. Will the next few decades show up to some future scientist as a time of great human change, brought about by new climates? We will see whether the volcanoes flare up, whether the sun's output changes, whether man's dust and carbon dioxide expand the westerlies.

We can't expect to control the forces that affect climate. We can't control volcanoes or the sun, we can't very well stop Third World people from farming their land and polluting the air, and we probably can't or won't stop the activity in our industrialized lands.

Perhaps, though, we can begin to realize the limitations that the earth, with its sequence of climates, places on our numbers and our actions.

REFERENCES

INDEX

References

Bach, W. 1976. Global air pollution and climatic change. *Reviews of Geophysics and Space Physics* 14(3):429–474.

Baerreis, D. A., and Bryson, R. A. 1966. Dating the Panhandle Aspect cultures. *Bulletin of the Oklahoma Anthropological Society* 14:105–116.

Bergthórsson, P. 1962. Preliminary notes on past climates of Iceland. Paper presented at the National Center for Atmospheric Research–Air Force Cambridge Research Laboratories Conference on the Climate of the 11th and 16th Centuries, Aspen, Colorado, June 16–24. (Unpublished mimeographed notes.)

Bergthórsson, P. 1969. An estimate of drift ice and temperature in Iceland in 1,000 years. *Jökull* (Journal of the Icelandic Glaciological Society) 19:94–101.

Blagden, C. 1781. On the heat of the water in the Gulf-Stream. *Philosophical Transactions of the Royal Society of London* 71: 334–344.

Borchert, J. R. 1950. Climate of the central North American grassland. *Annals of the Association of American Geographers* 40: 1–39.

Brinkmann, W. 1975. Unpublished chart. Center for Climatic Research, University of Wisconsin–Madison.

Broecker, W. A. 1968. In defense of the astronomical theory of glaciation. *Meteorological Monographs* 8(30):139–141.

Bryson, R. A. 1966. Airmasses, streamlines, and the boreal forest. *Geographical Bulletin* (Canada) 8:228–269.

Bryson, R. A. 1967. Is man changing the climate of the earth? *Saturday Review*, April 1.

Bryson, R. A. 1972. Climatic modification by air pollution. In Polunin, N., ed., *The Environmental Future*. London: Macmillan. Pp. 134–154.

Bryson, R. A. 1973. Drought in Sahelia: Who or what is to blame? *Ecologist* 3(10):366–371.

Bryson, R. A. 1974a. A perspective on climatic change. *Science* 184: 753–760.

Bryson, R. A. 1974b. Changing climates–changing times. *North Land* (Journal of Sigurd Olson Institute of Environmental Studies, Northland College, Ashland, Wisconsin) 2(2).

Bryson, R. A. 1974c. Heyuppskera: An heuristic model for hay yield in Iceland. Research Institute Neðri Ás, Hveragerði, Iceland. Bulletin 18.

Bryson, R. A. 1975a. The lessons of climatic history. *Environmental Conservation* 2(3):163–179.

Bryson, R. A. 1975b. Some cultural and economic consequences of climatic change. Institute for Environmental Studies, University of Wisconsin-Madison, Report 60.

Bryson, R. A., and Baerreis, D. A. 1967. Possibilities of major climatic modification and their implications: Northwest India, a case for study. *Bulletin of the American Meteorological Society* 48(3):136–142.

Bryson, R. A., and Baerreis, D. A. 1968. Climatic change and the Mill Creek culture of Iowa. *Journal of the Iowa Archaeological Society* 15–16:1–358.

Bryson, R. A., and Dittberner, G. J. 1976. A non-equilibrium model of hemispheric mean surface temperature. *Journal of the Atmospheric Sciences* 33(11):2094–2106.

Bryson, R. A., and Starr, T. 1976. Indications of Chandler compensation in the atmosphere. Paper presented at the Conference on Recent Climatic change and the Food Problems, October 1976, University of Tsukuba, Japan. (Proceedings to be published 1977.)

Bryson, R. A., Baerreis, D. A., and Wendland, W. M. 1970. The character of late-glacial and post-glacial climatic changes. In *Pleistocene and Recent Environments of the Central Great Plains.* Department of Geology, University of Kansas. Special publication 3.

Bryson, R. A., Irving, M. W., and Larsen, J. A. 1965. Radiocarbon and soils evidence of former forest in the southern Canadian tundra. *Science* 147:46–48.

Bryson, R. A., Lamb, H. H., and Donley, D. L. 1974. Drought and the decline of Mycenae. *Antiquity* 48:46–50.

Bryson, R. A., Ross, J. E., Hougas, R. W., and Engelbert, L. E. 1974. Climatic change and agricultural responses: A statement on research and technological priorities between now and the year

2000. A Report to the National Science Foundation. Institute for Environmental Studies, University of Wisconsin-Madison.

Carpenter, R. 1968. *Discontinuity in Greek Civilization*. New York: W. W. Norton.

Cess, R. D. 1976. Climate change: An appraisal of atmospheric feedback mechanisms employing zonal climatology. *Journal of the Atmospheric Sciences* 33(10):1831–1843.

Cumming, W. P., Skelton, R. A., and Quinn, D. B. 1971. *The Discovery of North America*. New York: American Heritage Press.

Das, P. K. 1968. *The Monsoons*. New Delhi: National Book Trust.

Donley, D. L. 1971. Analysis of the winter climatic pattern at the time of the Mycenaean decline. Unpublished Ph.D. dissertation, University of Wisconsin-Madison.

Du Bois, V. D. 1974. The drought in Niger. Part 1. The physical and economic consequences. New York: American Universities Field Staff Reports. West Africa series 15(4).

Du Bois, V. D. 1975. A note on the Sahel. New York: American Universities Field Staff Reports. West Africa series 16(4).

Dzerdzeevskij, B. L. 1968. Circulation mechanisms in the atmosphere of the northern hemisphere in the 20th Century. Institute of Geography, Soviet Academy of Sciences. Trans. (1970) by R. Goedecke for Center for Climatic Research, University of Wisconsin-Madison.

Ehrlich, P. 1974. *The End of Affluence*. New York: Ballantine.

Flohn, H. 1965. Probleme der Theoretischen Klimatologie. *Naturwissenschaftliche Rundschau* (Stuttgart) 18(10):385–392.

Fowler, M. L. 1975. A pre-Columbian urban center on the Mississippi. *Scientific American* 233(2):93–102.

Franklin, B. 1786. A letter from Dr. Benjamin Franklin, to Mr. Alphonsus le Roy, Member of several Academies, at Paris. Containing sundry Maritime Observations. *Transactions*, American Philosophical Society 2:330.

Franklin, B. 1834. *Memoirs of Benjamin Franklin*. Philadelphia: M'Carty and Davis. P. 373.

Fultz, D. 1961. Developments in controlled experiments on larger scale geophysical problems. *Advances in Geophysics* 7:1–103.

Fultz, D., et al. 1959. Studies of thermal convection in a rotating cylinder with some implications for large-scale atmospheric motions. *Meteorological Monographs* 4(21).

Gerster, G. 1975. River of sorrow, river of hope. *National Geographic* 148(2):167.

Haggard, H. W. 1929. *Devils, Drugs, and Doctors.* New York: Halcyon House.

Herrmann, P. 1954. *Conquest by Man.* Trans. by Michael Bullock. New York: Harper.

Hirschboeck, K. 1976. A new worldwide chronology of volcanic eruptions. Mimeographed. Center for Climatic Research, University of Wisconsin-Madison.

Kukla, G. J., and Kukla, H. J. 1974. Increased surface albedo in the Northern Hemisphere. *Science* 183:709–714.

Kutzbach, J. E., Bryson, R. A., and Shen, W. C. 1968. An evaluation of the thermal Rossby number in the Pleistocene. *Meteorological Monographs* 8(30):134–138.

Ladurie, E. L. 1971. *Times of Feast, Times of Famine: A History of Climate since the Year 1000.* Trans. by Barbara Bray. New York: Doubleday.

Lamb, H. H. 1966. *The Changing Climate.* London: Methuen.

Lamb, H. H. 1969. Climatic fluctuations. In Flohn, H., ed., *World Survey of Climatology.* New York: Elsevier. Pp. 173–249.

Lamb, H. H. 1970. Volcanic dust in the atmosphere; with a chronology and assessment of its meteorological significance. *Philosophical Transactions of the Royal Society of London: Mathematical and Physical Sciences* 266(1178):425–533.

Lamb, H. H. 1972. *Climate: Present, Past, and Future.* London: Methuen.

Libby, L. Forthcoming. Correlation of historic climate with historic prices and wages. *Indian Journal of Meteorology and Geophysics.*

Manabe, S., and Wetherald, R. T. 1967. Thermal equilibrium of the atmosphere with a given distribution of relative humidity. *Journal of Atmospheric Sciences* 24(3):241–259.

Manley, G. 1953. The mean temperature of central England, 1698–1952. *Quarterly Journal of the Royal Meteorological Society* 79(340):242–261.

Manley, G. 1974. Central England temperatures: monthly means 1659–1973. *Quarterly Journal of the Royal Meteorological Society* 100:389–405.

Martell, G. 1976. A climatic analysis of English wheat prices 1200 to the present. Unpublished research paper. Department of Meteorology, University of Wisconsin–Madison.

Milankovitch, M. 1920. *Théorie Mathématique des Phénomenes Thermiques Produits par la Radiation Solaire.* Paris: Gauthier Villars.

Milankovitch, M. 1930. Mathematische Klimalehre und astronomische Theorie der Klimaschwankungen. *Handbuch der Klimatologie* 1(a). Berlin: Gebrüder Borntraeger. Pp. 1–176.

Milankovitch, M. 1938. Astronomische Mittel zur Erforschung der erdgeschlichtlichen Klimate. *Handbuch der Geophysik* 9. Berlin: Gebrüder Borntraeger. Pp. 593–698.

Mitchell, J. M., Jr. 1963. On the world-wide pattern of secular temperature change. In *Changes of Climate*, Arid Zone Research 20. Paris: UNESCO. Pp. 161–181.

Moran, J. M. 1976. Glacial maximum tundra: A bioclimatic anomaly. Department of Geography, University of Illinois at Urbana-Champaign. Occasional paper 10.

Mulligan, H. A. 1975. Aid and rains win time for African's arid Sahel. Associated Press dispatch in *Milwaukee Journal*, April 27.

National Research Council of the National Academy of Sciences. 1975. *Understanding Climatic Change: A Program for Action*. Washington.

Pettersson, O. 1914. *Climatic Variations in Historic and Prehistoric Time*. Berlin: Springer. Reprint of *Svenska Hydrografisk-Biologiska Skriften*, vol. 5.

Rapp. A. 1974. A review of desertization in Africa: Water, vegetation, and man. Secretariat for International Ecology (Stockholm), Sweden, Report 1.

Schneider, S. H., and Mass, C. 1975. Volcanic dust, sunspots, and temperature trends. *Science* 190:741–746.

Scientific American. 1883. 49:296, 327, 352, 377, 389.

Singh, G. 1971. The Indus valley culture. *Archaeology and Physical Anthropology in Oceania*. 6(2):177–189.

Steensberg, A. 1951. Archaeological dating of climatic change in North Europe about A.D. 1300. *Nature* 168:672–674.

Stefansson, V., ed. 1938. *The Three Voyages of Martin Frobisher*. From 1578 text by George Best. London: Argonaut Press.

Stolle, H. J. 1975. Climatic change and the Gulf Stream. Mimeographed. Department of Meteorology, University of Wisconsin-Madison.

Swain, A. M. Forthcoming. History of vegetation and climate in northern Wisconsin, U.S.A., during the past 2,000 years. To be published in *Proceedings of International Union for Quaternary Research*, Tenth International Congress, August 1977.

Wahl, E. W. 1968. Comparison of the climate of the eastern United States during the 1830s with the current normals. *Monthly Weather Review* 96:73–82.

Wahl, E. W., and Lawson, T. L. 1970. The climate of the mid-nineteenth century United States compared to the current normals. *Monthly Weather Review* 98:259–265.

Watt, K. E. F. 1974. *The Titanic Effect*. Stamford, Conn.: Sinauer Associates.

Webb, T., and Bryson, R. A. 1972. The late- and post-glacial sequence of climatic events in Wisconsin and east-central Minnesota: Quantitative estimates derived from fossil pollen spectra by multivariate statistical analysis. *Quaternary Research* 2:70–115.

Wendland, W. M., and Bryson, R. A. 1974. Dating climatic episodes of the Holocene. *Quaternary Research* 4:9–24.

Wendland, W. M., and Donley, D. L. 1971. Radiocarbon calendar age relationship. *Earth and Planetary Science Letters* 11:135–139.

Wexler, H. 1952. Volcanoes and world climate. *Scientific American* 186(4):74–80.

Wright, H. E., Jr. 1968. Climatic change in Mycenaean Greece. *Antiquity* 42:123–127.

Index

DESIGNED BY ROBERT CHARLES SMITH
COMPOSED BY THE BLUE RIDGE GROUP, LTD.,
EAST FLAT ROCK, NORTH CAROLINA
MANUFACTURED BY GEORGE BANTA COMPANY, INC.,
MENASHA, WISCONSIN
TEXT IS SET IN PRESS ROMAN, DISPLAY LINES IN KABEL HEAVY
AND PRESS ROMAN

Library of Congress Cataloging in Publication Data

Bryson, Reid A
Climates of hunger.

Includes bibliographical references and index.
1. Climatic changes. 2. Paleoclimatology.
3. Droughts. I. Murray, Thomas J., 1943–
joint author. II. Title.
QC981.8.C5B77 551.6 76-53649
ISBN 0-299-07370-X